T0269961

Praise for *Digital Wellbeing*

"Caitlin Krause's *Digital Wellbeing* masterfully explores the intersection of technology and human potential. This book is a must-read for those who want to leverage the power of AI to unlock their creativity and imagination and ultimately invent new means of expression that will go way beyond human language, art, and science of today."

—Ray Kurzweil,
inventor, futurist, and author of New York Times bestseller *The Singularity Is Nearer*

"*Digital Wellbeing* is a welcome paradigm shift that invites us all toward a brighter future with tech. The book is brimming with compelling questions and practical insights— it is truly a resource to revisit. Krause is wonder-full as a guide. From her 'imagination index,' to 'the presence pyramid,' 'metaverse fluency,' and 'collective effervescence,' Krause gifts us with new language to open up important conversations about digital thriving."

—Dr. Emily Weinstein,
Co-Founder and Executive Director, Center for Digital Thriving, Harvard Graduate School of Education

"In today's rapidly evolving world, *Digital Wellbeing: Empowering Connection with Wonder and Imagination in the Age of AI* stands out as an essential guide for leaders and lifelong learners. This book skillfully addresses the challenges of modern life, offering timeless wellbeing principles that are both innovative and practical. It's an invaluable resource for anyone looking to thrive at the intersection of technology and personal growth, a must-read for those dedicated to enhancing their professional and personal lives with wonder, wisdom, and authenticity."

—Chip Conley,
New York Times bestselling author and Founder of Modern Elder Academy

"At a time when the impulse to lead fully digital lives and automate is tempting, Caitlin Krause's *Digital Wellbeing* offers ways to uphold creative freedom and imagination. She lights a way for integrity in this digital age, recognizing our lives as a Möbius strip of inner-outer connection. This book guides us to access what is most genuine and vital, fostering a mindful and authentic relationship with technology."

—Parker J. Palmer,
author of *Let Your Life Speak, A Hidden Wholeness,*
and *On the Brink of Everything*

"There's another plane of existence waiting for you, full of so much wonder and meaning. In this delightful book, Caitlin Krause opens your eyes to the marvels that can be found in the digital world. She's the perfect guide!"

—Scott Barry Kaufman, PhD,
author of *Transcend*

"Caitlin Krause is hands-down the 'go-to expert' on mental wellbeing and virtual reality. In addition to being a huge thinker on wellbeing in general, she has logged thousands of hours in-headset. She doesn't just talk the talk, she walks the walk. I can't think of a better person to write this book."

—Jeremy Bailenson,
founding director of Stanford University's
Virtual Human Interaction Lab

"Skateboarders become one with their boards to access everyday streets in novel, shared ways; that oneness blurs borders between board and terrain, body and mind, tunneling deeper into how we think and relate, as people. In a resonant way, new technologies are increasingly intertwining with tendrils of our senses, feeding the very perceptions that shape our conceptions of the world around us. Caitlin Krause maps this meshing of the human and digital domain in ways that elevate connection with others and reinvigorates a sense of wonder for the road ahead—all with such graceful clarity."

—Rodney Mullen,
skateboarder, fellow at MIT Media Lab

"Caitlin Krause's *Digital Wellbeing* is a transformative guide for today's leaders, showing how technology can enhance—rather than hinder—our capacity for genuine human connection and creativity. It's a vital resource for anyone who wants to create meaningful impact and navigate the emerging digital world with innovation and authenticity."

—Dorie Clark,
Wall Street Journal bestselling author of *The Long Game*
and executive education faculty, Columbia Business School

"The digital world offers us both challenges and opportunities that our wise and committed guide, Caitlin Krause, explores deeply in the comprehensive chapters of this practical and informed journey into human flourishing. *Digital Wellbeing* provides a feast of science-backed insights into how our attention, emotion, and social interactions are molded by an array of distractions and immersions, from work-focused email to social media. We learn about how the expansive dives into the virtual world and the expansion of our toolkits with AI offer exciting and concerning extensions to our digital lives. With a clear message, a compelling strategy for enhancing health, and illuminating examples illustrating how to maintain and regain our resilience, this book pulls us in and leaves us with important and memorable ways to thrive in the digital domain without become lost in it."

—Daniel J. Siegel, MD,
New York Times bestselling author of *Aware* and *Mind,*
Co-Founder Mindsight Institute
and the UCLA Mindful Awareness Research Center

"In this book, Ms. Krause serves as an imagination architect who identifies wonder and awe as essential ingredients to inspire a field of experience design that aims to fundamentally transform and support wellbeing. Her work elucidates the much-needed qualities of creativity, play and interconnection with elegance and simplicity, and invites us via a digital-lens to see the world real and imagined in a new light."

—Philippe Goldin, PhD,
Professor, University of California Davis

"In *Digital Wellbeing*, Caitlin Krause brilliantly explores how we can harness AI and digital technologies to enhance our physical and mental wellbeing. Her playful approach brings movement into the realm of discovery and awe, revolutionizing how we motivate ourselves to get up and moving. Krause's insights on using immersive experiences to boost body awareness are groundbreaking, offering a fresh perspective on thriving in our increasingly digital world."

—Kathy Smith,
fitness icon and *New York Times* bestselling author

"Caitlin is a world treasure. Every time I read what she has written or give her a hug I am a better person. Take to heart her message that wonder and imagination are not threatened by digital and AI, but instead, like other tools of our age, they can empower us, especially when we listen to and follow the promptings of our own human hearts."

—Tom Furness,
Professor Emeritus and "Grandfather of Virtual Reality"

"In this age of mass confusion and fear about the roles of immersive technologies and AI in our lives, *Digital Wellbeing* by Caitlin Krause distinguishes itself by blending practical frameworks with imaginative concepts to foster digital wellbeing. Unlike other books that focus on digital detox, Krause emphasizes integrating technology mindfully to enhance creativity, connection, and overall human flourishing. We need to embrace change to find the peace we seek, not hide from it."

—Alvin W. Graylin,
bestselling author of *Our Next Reality*
and global vice president at HTC

"We are each surfing an ocean of digital connections, and yet we generally feel less connected than ever. This book covers what we know and can study about digital wellbeing as an ever-growing aspect of our lives. The goal is not necessarily about maximizing profit, or productivity, or even the elusive state of 'flow,' but achieving what we all generally long for in a happy life: belonging, love, purpose, and real vs. fake connection. Caitlin Krause leads us along this journey with a book that is both informative and thought provoking and well worth your time to read."

—Avi Bar-Zeev

Pioneer in Spatial Computing; helping Google Earth,
MS HoloLens, Amazon Echo Frames, and Apple Vision Pro

"Caitlin's book is full of great insights into the interaction of psychology, technology, and wellbeing. A very informative text on how games can be used to make our lives better in meaningful ways that go beyond just entertainment."

—Noah Falstein,

Games for Healthcare expert,
former chief game designer at Google

"Caitlin's imaginative approach to the digital world is not only a gift for the present moment, where many of us are in need of reestablishing a healthy relationship with devices, but it is also an experiential peek at a version of the digital future where technology truly encourages human flourishing."

—Royce Branning,

Co-Founder, clearspace

"*Digital Wellbeing* by Caitlin Krause offers a window into creativity and wonder while defining and exploring tech and wellness. These topics often feel nebulous, but Krause provides a framework that readers can follow that will help them in their exploration of technology and how it can both inform and improve our lives. I highly encourage any reader to accept Krause's invitation to be lifted through digital wellbeing and all it encompasses."

—Starr Sackstein,

author and assessment reformer and disrupter

"Caitlin Krause has brought an insightful and transformative view to the intersection of technology and human flourishing with her book *Digital Wellbeing*. Caitlin's expertise and vision reflect her deep understanding of enhancing digital experiences with creativity and mindfulness. Her insights empower readers to cultivate a sense of wonder and imagination, driving personal and professional growth in the digital age. Her book encourages readers to act empathetically and positively impact others in their digital interactions. This book is a must-read. It will open your eyes to the profound importance and impact that the digital world has on all of us."

—Dr. Andy Clayton,
Assistant Professor of Leadership, Air University, Leadership and Innovation Institute

"*Digital Wellbeing: Empowering Connection with Wonder and Imagination in the Age of AI* by Caitlin Krause empowers leaders to blend technological innovation with genuine human connection, particularly those aiming to lead with purpose and compassion in our increasingly digital world. With this book, Krause has created a call to action for those ready to pioneer with strength and empathy, making it an indispensable resource for visionary leadership."

—Shelley Zalis,
Founder and CEO of The Female Quotient

"In *Digital Wellbeing*, Caitlin Krause has shown us that wellbeing begins with connectedness: to ourselves, to each other, and to the AI-enabled world around us. It's fundamental."

—Gary A. Bolles,
author, *The Next Rules of Work*

"Caitlin Krause's book is a masterclass in digital wellbeing. Our tech often overwhelms, and Krause shows us how to reclaim awe and wonder. Essential reading for anyone looking to succeed in this age of noise, disruption, and change."

—Scott Galloway,
Professor of Marketing, NYU Stern School of Business

DIGITAL WELLBEING

"Timeless wellbeing principles that are both innovative and practical ... an essential guide."
—Chip Conley

Caitlin Krause

DIGITAL WELLBEING

EMPOWERING CONNECTION WITH WONDER AND IMAGINATION IN THE AGE OF AI

FOREWORD BY MARC PRENSKY
INTERLUDE BY RODNEY MULLEN

WILEY

Published by John Wiley & Sons, Inc., Hoboken, New Jersey.
Published simultaneously in Canada.

For general information on our other products and services or for technical support, please contact our Customer Care Department within the United States at (800) 762-2974, outside the United States at (317) 572-3993 or fax (317) 572-4002.

Wiley also publishes its books in a variety of electronic formats. Some content that appears in print may not be available in electronic formats. For more information about Wiley products, visit our web site at www.wiley.com.

Library of Congress Cataloging-in-Publication Data is Available:

ISBN 9781394281787 (Cloth)
ISBN 9781394281794 (ePub)
ISBN 9781394281800 (ePDF)

Cover Design: Wiley
Cover Image: © art_of_sun / Adobe Stock
Author Photo: Courtesy of the Author

SKY10081033_080524

For everyone looking to ignite imagination
and bring wonder to life,
giving it wings

Contents

Foreword

What I've learned most from Caitlin Krause is that there is another plane to live on than the mundane and ordinary—and that with some guidance, we can all get there. Caitlin lives on that higher plane and wants to bring you, me, and everyone else onto it—all the time. She calls the plane "wellbeing"—a state of presence, observation, relaxation, low stress, awe, wonder, satisfaction, flow, and many other things.

How people can reach and spend more of their time on that plane—rather than in the more tedious and stressful parts of life—is the question and quest that consumes Caitlin. She is everywhere—speaking at TED or LinkedIn, giving workshops, holding meditations, and designing activities and games in real and virtual 3D worlds. She sees further into the future than many of us, and she understands that a new era is upon humankind—I call it "the third millennium"—and that in that new era people—and their activities—will be different and more evolved. As such she is a valuable guide.

As a guide, she has a lovely voice—my first thought was of Galadriel, the elf-queen from *Lord of the Rings,* talking to me, calming me down with her way of speaking. This book is a good reflection of who she is—open (one of her favorite words) to ideas and people. In the book she introduces us to many concepts and people she admires and helps us understand why.

The book is not just about wellbeing, but about *digital* wellbeing. It's about how we can all thrive, and become more evolved through incorporating the new digital world as a full part of us—while being human.

The digital part of humans is still burgeoning—it is less than a hundred years old in a human history of more than a hundred thousand years. But this new human evolutionary step (now nonbiological, like the language, social organization, writing and print, science and other evolutions since we became *homo sapiens*) is on an exponential trajectory that has passed the famous "elbow" of the curve and is now going almost straight up. Caitlin sees where this is going and how positive it will be for humanity because we will all, in the third millennium, get to live more of our lives on that higher plane.

Many of us are just now discovering that that plane exists. Many associate digital with harm and not the higher plane of wellbeing. But it won't be that way for long. The harms will gradually be reduced and the positive promises kept.

I think you will enjoy this book—I did. The journey goes from what it's like being on the higher plane to how we get there. It's filled with all the spaces and directions that Caitlin has explored and loves. Her aim is to not just describe, but to bring us there—and she offers many examples of wellbeing being produced and of models we can use to get more.

At its heart, this book is an invitation to open up our game and live, at least partially, at a new level, which—because of the exponentially growing compute power that defines the third millennium—is now available to more and more of us, and will soon be accessible to all. All of our lives are already digitally entwined—scary as that might seem to us. We all need to start learning to live our new digitally integrated lives in ways that are powerful, uplifting, and healing. Sadly, in the name of greed, some adults have already stepped in and corrupted parts of the digital world, allowing it to become harmful. But that doesn't have to be the way it goes forward—and it won't be if voices like Caitlin's are widely heard. Read on, and you, and those you know and influence, will be lifted—gently, with curiosity and agency—to a brighter and better way of being—a way of *digital wellbeing*.

—*Marc Prensky*

Preface

Discoveries and questions about digital wellbeing come alive in this book, as topics are approached through a new lens of wonder and imagination. It's been a *meta* sort of process, to adopt, reflect, and aim to embody this expansive approach to wellbeing while writing deeply about a topic that has so many transformative, astonishing—and at times misleading—facets and connotations.

How many of us, including our friends, families, partners, children, colleagues, and teams, need better ways to approach wellbeing right now? And how much does it feel as if tech is getting in the way instead of becoming the bridge to meaningful connection, presence, and imagination? This book changes that story. Here, as we demystify concepts, we give them greater context in ways that apply to our personal and professional lives, all while going beyond buzzwords and emphasizing the essential link between intention and attention. Our purpose in this book's journey is to elevate meaningful connection through experiences, actionable strategies, and dialogues that reach us right where we need it most, in support of full-life thriving.

Our focus involves AI because we're not just living in a digitally mediated time, we're in a transformative age where most of technology, including spatial computing, is powered by AI. It connects all of us and can serve to uplift us. It impacts how we lead, design, innovate, and implement better ways of working and connecting. The book dives into these deeper subjects, illuminating examples of how digital wellbeing is animated by wonder and imagination, within digital systems that use advanced technology as part of their infrastructure. Feeling connected and empowered, inside and out, as humans in this adventure together, is essential.

As we explore new digital applications for wellbeing, we are anticipating and meeting needs in a spatial computing landscape that will only grow. This moment is significant: humans are establishing new relationships with tech and discovering new ways of approaching ourselves and each other in this interconnected digital landscape. Wonder plays a significant role in our ability to thrive and to transform dreams into realities. This book explores all of this and more, as a guide and practical resource for personal and professional life.

As leaders of an emergent future, it's up to us to have a wider field of vision, which is what this book offers. The journey of it coming to life has been interesting and fascinating. I use that word, "interesting," which my fifth grade librarian told us should never *ever* be in a book report, but I choose it because it's appropriate. It *is* interesting, astonishing, and curiosity-inducing, the process of writing a book about wonder and wellbeing in the modern age, when wellbeing itself is a core metric of success in all arenas.

The interesting part this time around, with this particular topic, is the quality of my own mind, that inner landscape at a time that is incredibly noisy. I'm writing a book about digital wellbeing and wonder. I made a commitment in this process to embody imagination-infused digital wellbeing while I was writing about it because I do believe in that old adage that "you are what you eat" and that a product will inevitably reflect the emotional, psychological, and spiritual state of its creator. This book is also deeper than a product, of course.

As we talk about wellbeing, our own wellbeing is paramount. Secure our own safety belts first. And there's even more to it than that. I've found that striving for perfection with wellbeing and most other topics is not a realistic aim. What's more real, at least for me, is being open—one major theme of this book—really experiencing things as they are.

Two weeks before the due date, a friend asked me how writing the book was going. I said, "The vinegar tastes like vinegar." He understood right away. It's a parable from *The Tao of Pooh*, an allegory about the three major philosophies of China: Buddhism, Confucianism, and Taoism. In the story, Buddha, Confucius, and Lao Tse are gathered around a pot of vinegar. In this interpretation, Buddha tastes the vinegar first and says it tastes bitter because life is bitter and filled with suffering and attachment; his answer is to transcend life with its extremes and look to live the middle way. Confucius tastes the vinegar and says it tastes sour because life is sour and broken; his answer is to create rules to keep people in line. Lao Tse tastes the vinegar and says it's sweet because life is sweet when it is what it is, and if it tastes like vinegar, it's good that vinegar is vinegar. Life is life. Everything in life is naturally good if it remains true to its inner nature. So when vinegar is vinegar, it can make Lao Tse smile.

Benjamin Hoff, author of *The Tao of Pooh,* commented, "From the Taoist point of view, sourness and bitterness come from the interfering and unappreciative mind. Life itself, when understood and utilized for what it is, is sweet. That is the message of 'The Vinegar Tasters.'"[1]

I told my friend "vinegar tastes like vinegar" because writing a book, even a book about wonder and imagination, is a challenging process, and it feels intrinsically true to itself. From an organizational and content standpoint, there are always choices and new ways to approach topics that are naturally complex, layered, and multifaceted. My motivation is to write something that is good for the world and useful, offering agency and ideas at a time when it feels not only nice to have but necessary. I cannot prescribe what I share; I can only offer it. My own imagination has felt as if it's brimming over with ideas related to digital wellbeing. This book is motivated by that sense of possibility and wonder, plus a deep conviction that this book is very much needed at this time as a way to inspire connection, creativity, and freedom over loneliness and sense of separation and disempowerment. Each detail I share will likely shift in the coming months when new research and views uncover more truths in this emergent intersectional field. Change is constant, and that is wonderful. This text becomes a part of the conversation, not static but adaptive and contextual. Use it as a sounding board.

This book contains related topics that are meant to be seen as "beads on a necklace." Each bead can be appreciated and approached as a singular object, and each is also related to the ones beside it. A thread connects each bead, stringing them together. As you read, it's possible to skip around, out of order, and still get a sense of the wholeness of the necklace. The message doesn't change. The beads could potentially be reorganized. They are in line in an order that I think is fitting, and yet each has its own genuine individuality.

In each chapter, each bead, of this book, I will place their meaning and relationship to digital wellbeing and wonder in context, and I will give you practical approaches and ideas for using the topical insights and knowledge to enhance your own life and inform your work practices. Many strategic applications and methodologies are useful for leaders and teams. The way you approach it is completely up to you, providing an intentional freedom. See what comes up for you. View this book as a series of conversations, where each bead also contains a world within. They are spherical and fully dimensional. While related to the whole, they have individual integrity, just like each of us.

At the end of each chapter, I've included several key reflection questions that will guide conversations and enhance your own sense of wonder about the topics. If you would like to keep a journal, it can become a helpful accompaniment to this book, as an exercise to stretch thinking and offer more than habit change.

Right now, as I'm writing this, I'm outside because the sun is shining and I like to write outdoors. To my left, a two-lane road with traffic going a steady pace speeds by, becoming a hum. To my right, a jackhammer is performing some staccato march in a nearby yard. Birds, mostly crows and songbirds, are varying caws and trills. Squirrels zipper up and down trees that have not yet grown leaves but are budding and signaling springtime. The sun shines warm on my back, and everything becomes part of the music of this book.

The listening mode is hard, given the noise of the world. At times "it's too much with us," isn't it? Yet maybe that's also what it means to be alive at this stage of time, to embrace that semi-chaos and try to find our own ways of being that align with openness and inherent truth.

Writing about digital wellbeing means that I'm listening to all sides. There are many sides, and it's not either/or; it's facets. There are "digital literacies" in pluralism, not singular. Wellbeing in a digital sense is universal because technology is embedded and extended in our way of being now, so it's about literacies to be able to do even more than live lives of comfort, but to live in ways that feel open and true and whole.

We will talk about that wholeness through the course of this book. Since this involves an emergent and complex set of topics that exists in the context of what is also adaptive and multifaceted, there are many extensions to explore. At the end of each chapter, I mention extensions I would like to investigate and develop in the future. Think of it as an emergent, generative, responsive map. I include a few points about where I see the conversation growing. If you would like it to expand, let me know, and there will be more where this came from.

By now, you might glean a little of my personality. I'm open to us getting to know each other better through the course of this book. In reverence to ritual, to cadence, and to finding what feels good, I'd like to name a few routines that allowed this book to come to life in the way that it did. As you read, these practices might be helpful for you too:

- Waking up to sunlight—stepping outside at least for a few minutes each morning. Even cloudy days offer some sunlight.
- Ideas first, phone later—I always get more creative work done when I work away from my phone.
- Deep work surges—writing while turning everything, including internet access, off, then going back to reference check and research and respond to messages.
- Exercise outdoors—as a break from work, as a release, as reflection, time to "be" and to feel my physical body exerting itself and enjoying nature.
- Getting enough sleep—"enough" varies by the individual. We should not feel guilty for getting eight hours. Most adults need about eight hours, and the worldwide average is 6.8 hours. It affects our baseline health in every sense.
- Declining some things to create space—the things that clamor for our attention will still pass by if we let them go. I think about things from five years ago that seemed to matter so much. And I think about my future self at 100 years old. What would she tell me for advice? And what would I want to look back on and remember?
- Saying "yes" to some things—paying attention to cues like visceral lifts, and whether I feel it would be good for the writing of the book—are a priority. I am curious and have a tendency to lean into a lot. I'm learning to create space between the notes and nodes.
- Practicing reappraisal and the art of reframing—it can help, especially in a world where things are beyond our control. In the past month, I have experienced natural uncertainties, including lost luggage the day before I was giving a keynote overseas, and it was this process of reframing that kept me stable and operating with some form of levity even while experiencing jetlag and stress.

- Neck stretches, finger stretches, body stretches—stretching in general. Taking deliberate breaks for movement doesn't have to mean all-out intense exercise sessions. Just spending five minutes stretching and walking around every 30 minutes or so keeps the flow.
- Writing to music—there are two types of people. Those who love to work and write to music, and those who are no fun. Just kidding. I write to music. I make soundtracks. Most of my creative writing music is wordless. I appreciate when friends share their music with me. I'm grateful for a music, movement, and meditation retreat I was part of at the start of this book-writing process. That music formed the score for the start of this book. Writing is a musical process. Even my percussive typing on keys feels like a musical endeavor, especially when I'm in flow state. It's not easy, but it feels true, and I'm grateful for that and for being able to be some form of conduit to share this with others, including you right now.
- Giving in to the good, noticing moments of wonder. It's rainy at this time of year. I have seen more rainbows in the past week than in the past five years combined. I don't know why. Maybe I'm looking up at the sky in between writing. Or maybe the writing about wellbeing is prompting me to notice the wonder and awe all around me. Maybe a combination of all of it. Maybe there is just a lot of rain and sun and consequently rainbows right now. Point is, I'm here, I'm alive, and I'm noticing and appreciating it all.

Let's enjoy the wonder and discovery of this book. Vinegar tastes sweet, doesn't it?

Introduction: Open Up to a World of Wonder

How the Imagination Opens the Mind

Just as mindfulness, wonder, and awe can open the mind and heart to adopting better approaches to digitally infused wellbeing, technology can become a tool to enhance mindfulness, creativity, wonder, and wellbeing. It's reciprocal and cyclical.

I'd like to invite you to engage with the following imaginative exercise, as you feel most comfortable. If closing your eyes creates discomfort, you could keep your eyes open:

Imagine a time in your life that made you feel completely inspired. You might have felt both big and small all at once, an extreme sense of delight, joy, and connection, or a loss of attachment to time and individual identity. Take a deep breath in and softly exhale, even more slowly than you inhaled. Now close your eyes and keep noticing the details of the experience you imagined, taking in all the sensations, the feelings of what that was like for you. When you're ready, open your eyes.

I have a few questions for us to reflect on, while this experience is fresh:

How would you describe the place you chose?
Was some sort of nature surrounding you?
Was there a feeling of vastness?
Were you surrounded by others, or were you by yourself?

What you just recalled was an awe-inspired experience. It might have been from childhood or your adult life. (If the two distinctions are one and the same, you're unlike most and have likely lived a beautiful life where childhood persists.) You could have been in a group or completely peacefully solitary. It could have happened in any time, in any place. One of the interesting parts is that, as 3D creatures, our awe-inspired moments tend to be deeply grounded in a sense of place. There's a maxim that "a place is a space with meaning" and that lands true for me. Environment matters.

This book is all about wellbeing in layers that start with our human physicality and our consciousness. These days, digital technology permeates our daily lives, enabling us to connect in meaningful ways, if we approach layers of technology mindfully. We might have "offline time" or engage in tech sabbaticals and retreats, but we inevitably return to a world that is connected and supported by technology. While we survey different approaches to wellbeing in this book and investigate the etymology of the term, the premise is that wellbeing and digital wellbeing are one and the same because of that layered approach. They have to do with openness—an open way of being in connection, using imagination and wonder to elevate and thrive. Context will change, and conditions will change, and I cannot tell you what it means to flourish in a way that is formulaic or conditional, in an "if, then" predictive model. We can share about wellbeing as an openness to experience, though, and that openness inevitably leads us back to connection with the senses, and an expansive mindset using wonder and the power of our imaginations to inspire us.

Think of the imaginative exercise you just did. It was an exercise where you engaged with a moment of wonder. These moments tend to have certain things in common. We are at ease with our surroundings, with ourselves, enlivened by them. Some people say these awe experiences "make them come alive." We might have felt, as mentioned, very big and very small all at once. We dropped into a different state, a lucid state where it was as if we were dreaming and waking all at once. Things came easily to us in these moments, which seemed completely detached from the normal constraints of time. We lost our inhibitions, even our sense of self-consciousness. We lost attachment to identity and all the extra stories we were carrying. There was a joy in this time, a lightness of movement. A oneness with ourselves and everything around us. Everything was in complete flow.

We as humans can have many awe-inspiring experiences once we start to notice the possibility for them and prioritize them. It all starts with our imagination and our willingness to surrender to these experiences with wonder that can be so transporting. Some people will say these are the best moments of life, the ones in which we are filled with awe and rise above what might seem limiting or constraining. We feel deeply connected, and also elevated.

As an imagination architect, I study wonder and awe and apply these principles to experience design in the most transformative ways.

If you've been a part of one of my talks, courses, or workshops, you may have been to a virtual garden space and introduced to an onboarding experience for virtual reality using objects as teaching metaphors for memory palaces. I have taken a group of government officials to the moon and led them in collaborative exercises. I've gone deep sea diving with a team of educators and watched them use the SCUBA method (mentioned in our virtual reality experience design and Hero's Journey models later in the book) in their design strategies for teaching and mentoring young people. I've created snowy landscapes with the northern lights and led poetry readings, and I've taken corporate teams hiking through the Alps and created expressive interactive art on the mountaintops.

Introducing ideas involving team collaboration and "shadow storytelling," we've gathered in caves where vines begin to grow from the cracks and meditative music guides people in their sharing of stories. I've built a moon garden for meditations and a walking path for shared exercises. Some experiences are meditations, and others are active collaborative sessions that lead a group in experiences with shared trust, emotional intelligence, empathy, resilience, and teamwork. There have been movement workshops, interactive design exercises, and all sorts of surprising nuances that add elements of mystery and delight. The best parts are the surprises, internal and external, so I invite you to try it for yourself sometime. Imagine becoming a ball of light and floating through the cosmos. It sounds trippy, I admit, but there's more to it. In "surreality," in a spatial world, this becomes a deeply intimate shared experience where new insights can and do emerge.

As this book is a series of "beads on a necklace" for us to consider and explore, we're incorporating awe, wonder, and imaginative delight as our digital wellbeing threads, or conduits. Wonder becomes a vehicle for transformation. We can access the wonder in ways that are technological as well as biological. This is about layers. When our approach is imaginative and open, new discoveries will emerge on an individual level. I'm inviting you to explore through a new digital lens, and I'll give you a deeper understanding about the tech tools as we go.

As you can imagine, I've designed experiences and led tours across time and space, in physicality, virtuality, and all instances of the full continuum of spatial computing, for decades. I've seen hardware and programs change and change again. It's not about the hardware. There's something else, and that is essentially one of the deep discoveries that this book is about.

The beauty is, too, that a lot of this involves experimentation and play. To design for wonder, you need to live wonder as a key ingredient. This is not about paint-by-number and forced outcomes. It's about cosensing each other, and it's about practicing new ways of engaging in community rhythms. Our communities are deeply missing rituals, especially with the global pandemic having disrupted our traditional ways to physically gather and share.

This book is essentially about connection. There are bright spots: This period of deep disruption and unrest did push us to find new ways to connect and to give some digital connection tools a chance in ways that were different and new. The challenge is that we might not have paid attention to the art of integration in mindful and intentional ways. We might have thought that this brave new world of technology for connection meant having endless zoom meetings all day, sitting still in a chair, and then watching television with that simultaneous mindless scrolling that Esther Perel talks about.

Autopilot mindlessness is not helping us, and the reverb back in response, now that the shock of the pandemic is behind us, could be to push harder against the opportunities of transformative technologies. It's time we took a collective breath after going though a pandemic together as a society. Many of us endured what was a numbing, unsettling, shocking, and lonely experience. Naming that is part of the healing.

For many of us, this time of disruption caused by a global pandemic was a time of loss and isolation, when hope battled despair. We found new ways to endure. I think about that word, *endure*, and what it means, and how tired we can be underneath it all. For those of us with others depending on us, as leaders, parents, teachers, friends in community, to be there for each other is collective resilience, and it can be individually and collectively exhausting to endure. How do we now stay supple and kind to ourselves and reappraise our vulnerability and all we are going through as a strength?

I say this, and mention *reappraisal*, the art of changing how we view a situation (even one in the past!) to shift its emotional impact, because a lot of people have felt shame over the past years, for how much is beyond our individual control, and we're at a loss to manage our own feelings. We lose hope and dive deeper into isolation and depression. I have found that "the only way through is through," with a gentle self-compassion, and it's with authenticity that we can truly look at our experiences and feel whole. I think about my own practices in gravity and levity, in naming my emotions and also encouraging play. I think of curiosity and willingness to be seen and see others. I think of the power of kindness, the messy, daring, brave kindness that can lead us through and give us a sense of strength and sanity. Brené Brown and others have promising research on the power of reappraisal and reframing,[1] and I use this science to offer what I consider a valuable resource: personal and collective freedom.

The research on reappraisal, particularly in the context of emotional regulation and vulnerability, emphasizes the power of reframing our emotional experiences and the importance of social connections in this process.[2] Cognitive reappraisal is an emotional regulation strategy that involves changing the way we think about a situation to alter its emotional impact. This approach aligns with the strength of connecting with vulnerability, courage, and the benefits of embracing emotional openness and social support. Cognitive reappraisal encourages individuals to actively reinterpret challenging situations in a more positive or neutral light, thereby reducing negative emotional responses and enhancing wellbeing.[3]

When people see their vulnerability as weakness, they are more likely to hide it. What we hide and conceal breeds shame. We might not intentionally do this, but it happens quite naturally, and then it disempowers us because we feel that we are somehow leading hidden lives. We sense the disconnection and the lack of completion, and we go about our days in worlds of concealment, worlds of survival where we are masking our pain and our brokenness instead of owning it and practicing compassion for ourselves, our loved ones, and our teams and communities.

Let's change this story, right here and now.

The brokenness is beautiful and natural. As Leonard Cohen says, *that's how the light gets in*. There are art forms of *Kintsugi* and *Kintsukuroi* dedicated to this, and entire philosophies of *wabi sabi* explain why the most powerful and beautiful experiences involve imperfection, breaking, and lack of pure prediction. There's a spontaneity to life and a responsiveness to the moment, and this impulse is blocked if we hold too tightly to perfection and prescription.

We do this to ourselves all the time, and the worst part is that our connected society often tightens that grip of "expectation" making us feel as if perfection and conformity are the goals, when in essence they are not.

As John O'Donohue so eloquently writes about, the goal is to engage with the wildness in our hearts. We cannot do that if we are listening to external conventions and prescriptions. What comes from us is our source and our wellspring. Parker Palmer calls it a "hidden wholeness" when we come to discover and live that undivided life that reflects our integrity.

The incredible part, which many might not expect, is that digital wellbeing experiences and tools can help lead us there and unblock and uncover these purest parts of ourselves, the sweetness and the wholeness. If we are consistent and accurate with our measures, true to our values, honest with our integrations, and mindful of our own bodies' ability to teach us, we can have a future where technology is helping to enhance our life instead of blocking it. This is what I've experienced firsthand, and it's deeper than a hope. It's a conviction that there are ways we can do this. Yet we must be deliberate and considerate at this stage, as technology affords us so much power, and we need to harness wisdom to know how to wield it.

As the Center for Humane Technology's Tristan Harris, Aza Raskin, and others will point out, Apple's original vision was that technology could be "a bicycle for the mind," allowing us to travel and experience new freedoms and joy. This was inspiring to more than a generation. That vision sparked a movement. Right now, many would argue, or even despair, that we are not on that track.

Yet, we might be looking through only one lens, or at a certain facet that reflects our own expectations and life experience. The answer is not to suffer through our own disenchantment, nor is it to restrict ourselves, our friends, and our collective children as a society from experiencing what is our freedom. Technology can and should be about freedom. To look only at harms and dangers is to rob ourselves of opportunity and rights to have agency in shaping the future.

Our future is undoubtedly going to continue to involve technology in ways that are expansive. Artificial intelligence (AI) is not going away. In fact, we will be creating new contexts for it and laying the foundation for how our society is designed with AI infrastructures and integrations in place. These early days of AI are a pivotal time for intention setting and making sure that a foundation is in place with the right vision, values, and benchmarks. How do we chart a course, and also course correct when we need to, if we don't have alignment at the beginning?

Right now, spatial computing technology is allowing us to share experiences like never before and to access new ways of perception, to think and feel and express in ways that tap into our imaginations and better equip us to respond to epiphanies that will undoubtedly spring forth. It's a world of promise, where these tools and interfaces invite our identities (yes, multiple, because we contain multitudes!) to express themselves and offer us opportunities for games and experiences to be even more engaging, even more empowering, and even more uplifting in a way that leaves us feeling connected, elevated, and inspired rather than depleted.

If we are undoubtedly integrating technology, why not design the integration correctly, with the best values and measures in place? Essentially, this is about wellbeing at the foundation. And digital wellbeing is the "how" of that technology integration with wellbeing in top priority. It is a vital cause right now that will determine the future of humanity.

It's up to us, and it's up to the "how" of digital wellbeing that empowers, connects, uplifts, and extends us.

Digital Wellbeing: An Extension of Us

There's a reason that XR stands for "extended reality." Digital is an extension of our selves, and it's a means by which anything is possible. Even take the word "digital" itself. The root of the word was originally tied to digits of the hands, once used as a physical means of counting, transferred over to systems of computing. Our hands used to do all the work; now, our "digits" are extended through computers.

The hands used to be the means by which we could execute our will. (Side note: I once wrote a paper in university that examined all of the Shakespearian characters and their expression of free (or coerced) will, as evidenced by their hands and the actions they committed by hand. My argument was that you could judge characters better by the hand than you could by eloquent soliloquies. When you zero in on hands, it becomes fascinating. From Lady Macbeth to Titus Andronicus, from Lear to Laertes, looking at hands is quite revealing.) Now, when we use digital technology, we create many extensions of ourselves, and anything and everything is possible. The worlds we help to create become as vast as our imagination.

We're establishing new metrics, new KPIs (key performance indicators) in a sense, for what it means to thrive in all aspects of life, including personal and professional. There is a possibility for wholeness and freedom, in elevating the way that we approach everything from business to education to health and relationships. The biggest determinant, then, will be our wellbeing, which is therefore reliant on the quality of our imaginative capabilities. How do we better prioritize for that? That very question is what this book is all about. This book considers the questions about embodiment and 3D technology at exactly a time when we are moving into spatialized modalities as primary means of connection. There are so many ways that digital wellbeing and spatial technology can create a layered life enhancement. Intentionally, it does not and should not replace the quality of physical connection.

Wellbeing of the future is much more than alleviating pain and preventing injury. It does more than heal; it uplifts. Our future is going to be one where we undoubtedly continue to interact and integrate with machines. Our purpose is no longer repeatable processes, and our minds are not best used for rote memorization. We need to cultivate wellness through creative imagination, play, and wonder. This is what humans are made for, and this is how we lead toward the future. Every major index, including the World Economic Forum's Future of Work skills, is calling out for imagination as a top skill. It is the creative mind that is going to lead the way. Our creative thinking is all about wonder. So, how do we increase plasticity and start to build these skills?

The answer lies in the sense of wonder, the art of play, and our future life and the lives of our future ancestors depend on it.

So, what does this entail? How do we prioritize and design for it?

There is no easy formula for creating digital wellbeing conditions that inspire connection, belonging, and ignite imagination through wonder and awe, though there are certainly elements that are necessary and habitudes to avoid at all costs. Wonder and imagination rarely happen purely in isolation. There's some ebb and flow; there's inspiration, reflection, and integration.

There are several paradoxes, and you'll find some here—these will be vital for our discussions throughout this book and beyond.

Many of us would like to be in flow state all the time, and yet it's also a cycle that can't be easily drilled into formula, try as we might. We also need downtime, to refuel, to rest, and to integrate learnings. There's no exact recipe, but the more self-awareness we have the more likely we are to ask for what we need and to open up to flow.

Being "in flow" involves openness and wellbeing essentially, and this paves the way for epiphany. Our great epiphanies can happen in any place, and in any time. In many ways, art is a translation of those epiphanies. Our ability to imprint and recall could be one of the ways that spatial environments and 3D tools can help aide us in the paths of discovery.

We can design for wonder and awe, and these are often fleeting experiences in the brain. What happens *before* and *after* could be the keystones and how we set ourselves up for success.

Older forms of leadership are being disrupted. Stoicism is no longer a strength. Suppleness, vulnerability, openness, and wonder lead to true thriving as a leader.

We work on teams, in collaborative ways, and we need to purposefully and deliberately defend and uplift wonder and awe. We need to design wonder-rich moments that allow us to see in new ways. This often takes great vulnerability because we are encouraging ourselves and colleagues to step out of comfort zones, to leave inhibitions behind.

The teams who are the strongest and the most innovative—the ones who are quicker to learn from mistakes and also take risks that lead to new discoveries—are ones who have created psychologically safe dynamics. Having psychological safety is key, and there are essential qualities that help as primers for wonder and awe.

Three Frameworks Guiding Our Digital Wellbeing Journey

Throughout this book, we will look at **three main frameworks** that help as guides when we're designing for wellbeing:

The first framework is the **Four Culture Cornerstones**. Everything is about relationships, especially our human-human relationships mediated by technology, and the way we approach setting the foundation makes a difference. These are cultural cornerstones.

The second framework or methodology is the **11 Keys to Imagination**. These are quality points that build an imagination-rich experience.

The third involves spatial design, since we are spatial creatures and spatial computing can and should be an imagination-rich extension of ourselves. These **Seven ThEmes of Spatial Design** all happen to begin with the letter E and are part of contributing to what makes a spatial experience special, filled with wonder and awe. Spatial computing is the way of the future, so mastering these Seven ThEmes ensures that we keep our own wellbeing and our values front and center.

It's vitally important to keep all three of these frameworks in mind as our benchmarks and guideposts, especially as we lead initiatives, foster creative thinking, and work with teams to encourage and empower transformation. They are essential learning and thriving frameworks for all ages. We will explore all of them deeply in Chapter Two.

Some people call the full continuum of spatial computing "Extended Reality" or XR. This is a great way to put it because spatial technology is an extension of ourselves. It can execute our will, in a sense. The key, or one of the keys, will be for future humans to decide what their wills will be. What do we actually want? We will have more power than we can dream about at this stage. It's starting to become clear to us, this potential.

So this is a good thing, the potential of digital technology to open our minds and hearts and allow us to extend ourselves, to reach others, and to also reach inside and better access our own imagination. We will address "why social matters" in this book and how we are social animals who are built to connect. We are wired that way. This is not about being an introvert or extrovert or ambivert (my supposed type), it's about knowing that whatever our type is, we all want and need to belong.

In the future, if not already, AI-powered computing systems, including robots, will be integrated in that sense of belonging, and awe and specifically the collective effervescence of sharing with others will all be wrapped up in that social context of community, contributing to it. The loneliness epidemic will diminish because we will prioritize connection instead of engagement in our designs.

The time is now to be informed about potentials and to make choices that empower the collective, not diminish their agency. That is essentially what this book is about.

By definition, a book that's about "opening up" is living the message of open curiosity. There are reflections along the way for you the reader to respond to and a lot of interplay of wonder as we posit possibility together. The best scientific findings are shared, and every view takes into account that times are shifting and new discoveries are being made by the minute.

It's a time of rapid change, so we must also mirror that agility and resilience in staying open with our thinking. We not only look to mentor younger generations, we look to them to teach us and show the way. Every time we feel fear rising up, which is a common reaction to technology discussions, we can invite ourselves to use a phrase, *"That's interesting,"* as we name the experience. If we are not unsettled by some of this, then we're not really engaging and grappling with it. As Gertrude Stein said, "I salute you and I say I am not displeased I am not pleased, | I am not pleased I am not displeased." This doesn't mean we need to be pessimistic or complaining, it just means that there's something at stake and we care. We open up to paradoxes, weigh opposites, and know that there is space between, and areas of convergence. Naturally.

What is digital wellbeing?
It's opening up. It's a quality of open.
A way of being.
The word "open" will stay with us throughout the book.

We have always lived in two worlds: the world of the mind and the world of our imagination. Now, there is also a *third space*: the worlds in shared virtual spaces, which are empowered by digital technology. Our worlds, and our sense of agency and thriving, can be extended, expanded, and elevated in connection.

They can all become spaces and places where wonder comes alive and where we experience digital wellbeing by design with imagination.

Worlds open, as do we. . . .

Caitlin Krause

1 Defining Wellbeing Through a New Lens

What Is Digital Wellbeing, Truly?

In my view, everything about the term "digital wellbeing" can be related to choice and agency. It involves choosing how to live our best, most fully expansive and inspired lives in a necessarily digitally mediated world. Digital is not going away, and the answer is not escapism. Rising to a different plane, a new way of being, thinking, and exchanging with others, is what it's all about. This "way" of wellbeing, using imagination and wonder as vehicles, is what this book is about, and it's about mindset before skillset and toolset. It involves transcendence, in a way that essentially connects us and still maintains our essential connection to humanity.

Prescriptions about wellbeing, and paint-by-number guides, are not what this book is about because those strategies tend to be short-lived, impersonal, and flawed in assuming that humans are robots devoid of context, emotion, and meaning. Instead, this guide to the art of being well in a digital world provides an essential foundation that will allow you to understand wellbeing first, apply it to the context of your life, and then show you ways to reach that higher plane.

In this book, I give you the ability to have a solid foundation surrounding the future of wellbeing, which will by definition and necessity be a way of digital wellbeing. You are the leaders who will be helping to shape this future, and you have everything you need, right here, right now. Each of us is equipped with our greatest asset to approach the emergent future, exponentially ripe with possibility: our imaginations!

This book invites you to tap into that imagination and sense of wonder even more deeply, and shows examples of what we can experience using digital wellbeing for connection, wonder, and joy instead of separation and loneliness. Through this text, you'll learn about the power of intention and attention over mindlessness, and you'll take away many ways to live a fuller, brighter "right now", which changes the quality of each moment, each day.

The "wellbeing" all humans need—whether we are old, young, working or in leadership or any other discipline—involves *spending more of our lives on a different level*—not completely away from the tasks of life that we all have to do, but more in the realm of *presence, imagination, wonder, awe, relaxation, play, openness, and flow*. Having more of this "wellbeing" in our lives benefits everyone—leaders, workers, students, guides, caregivers, humans of all ages and stages. When you have access to wellbeing, more gets done, and it is of a quality that is truer to our sense of purpose, presence, and intention. We become creatures living our lives in that space of deep connection, which can provide the greatest fulfillment.

All of us have access to this wellbeing plane inside of us, but most of us rarely see or experience it, and we lack the means to bring it out in ways that are useful and valued. Martha Graham, John O'Donohue, and many other leaders have spoken about what happens when we stifle our own wonder and creativity, and how fear can prevent us from taking risks. Just imagine what epiphanies—what pure feats of imagination—have been completely lost throughout history because there was never a path for us to get it out and because we were afraid of doing so. As O'Donohue says, "Fear is negative wonder." I am here to acknowledge the fears, name them, and add wonder to the equation in ways that mitigate fear and offer practical freedom and connection.

We all want to share more connected lives, with many avenues for expression and thriving, and this book is about the ways to amplify and act on that intention. Digital is a conduit; it's all media. The human-to-human, and human-to-nature, and essentially human-to-epiphany/wonder/awe/

flow can be mediated by technology. Tech can become a bridge, in a similar way to that of using any vehicle or substance to provide an onramp to a higher plane. Our human experience is still the focus, and we guide the pathways.

Until now—i.e., before digital—there has never been a path to access the higher plane with such reliability, agency, autonomy, and clarity. Now there is—and it's called a virtual experience, or spatial computing. It's not substance-dependent, and it does not necessarily involve any prescriptive spiritualism. It's a newly accessible part of human reality, and it's a path to accessing the part of yourself that you have and need, but typically do not use. It's called "wellbeing," and I am a guide to help you—and everyone you know and collaborate with, in professional and personal relationships—access it to your benefits. This book is your guide to finding this wonder-rich digital path to wellbeing—a part of you that is necessary to be successful at anything—leading, working, parenting, evolving, nurturing, and thriving—in this modern digital age.

> *There is a vitality, a life force, a quickening that is translated through you into action, and there is only one of you in all time, this expression is unique, and if you block it, it will never exist through any other medium; and be lost. The world will not have it.*
>
> —*Martha Graham*

> *So many people are frightened by the wonder of their own presence. . . . Many of us get very afraid and we eventually compromise. We settle for something that is safe, rather than engaging the danger and the wildness that is in our own hearts. . . .*
>
> *Fear is negative wonder.*
>
> —*John O'Donohue*

The Art of Listening and Being Heard

Around the world, as far as I've experienced, the standard greeting, "How are you?" is typically answered with a customary response: "I'm fine" or "Doing well," or something of that general nature. Regardless of the choice of phrase, we aren't expected to give a complex answer, and most people aren't listening for the qualitative content of the response. We might, if we are lucky, be tuning into each other's tone and emotion. We generally volley back the same question and are given similar answers. We're typically already moving on with our day by the time the response reaches us. How would we rate our listening here? Leaders will say listening is the number one skill. How do we practice it with intention?

"How are you doing?" has morphed in our global culture into a mutual acknowledgment of presence rather than an invitation for deeper conversation. We rarely attend to each other's responses in a way that would show care. When we respond that we are *doing well*, what does that actually mean, and is it the truth of our experience?

When I recently asked a friend how she was doing, she took time to thoughtfully consider and answer. It was a connected moment. She pointed out that she's choosing to respond to how she is doing *right now in this moment* rather than give a broader general answer. I appreciate that. There are many ways to answer a question about how we are doing, and it should give us pause. We need space between the notes, to greet each other, see each other, and practice the essential art of listening.

How often do we each take time to listen, connect, and share how we are, and do we approach it in a macro sense or a moment-by-moment outlook? To let each other know about our wellbeing is a mutual sort of consent and trust. What is sometimes viewed as an individual state can also be addressed systemically, at a community level and a societal level. Wellbeing is a complex and interrelated concept.

What's in a Word?

When it comes to being well, and the changing states of wellbeing, the same can also be true about micro and macro states of time. How we are—if we are "well," to varying degrees—can and will change many times, even over the course of a single day. It's not static, and it's very context and condition dependent.

I choose to bring all of this up because I actively question terms and often look to define them by contexts and use cases rather than connotations. You'll see that sort of approach throughout this book. Defining by overtones can get us into trouble because we can easily limit a word to its subjective associations, which are often its most commercial exposures rather than its true intentional meaning.

The meaning often lies in experience, and this book aims to investigate experiences, facets, and angles that illuminate all of the dimensions of wellbeing. It can be liberating rather than limiting, invitational rather than prescriptive, and encouraging rather than shame-inducing.

We can feel the brokenness in our world if we look closely enough. It's not hard to see the pain, separation, destruction, and malaise. It's also easy, once we pause to notice, to find the exquisite beauty, the care, the joyful possibility, and the wonder that moves us to astonishment.

When we investigate wellbeing, let us detach from judgment and allow ourselves to simply notice, to get curious and even playful with topics that can be intimidating and heavy. It does not mean we take the topic lightly, but we can carry lightness in this process of investigation and learning.

We then find out even more by seeking new understandings, blending our curiosity with ways to use life experience, which encompasses our emotional states, memories, previous learnings, body awareness, and behaviors.

I'm conscious that I do not have all of the answers, and wellbeing is a term with different interpretations. It can be seen as a buzzword associated with pop culture, business, academia, medicine, self-help, and psychology. The same can be said for mindfulness, wonder, and awe, creative fields in which I've worked for many years. Issues can quickly become polarized, with people using specific words and lingo to define tribes of like-mindedness.

While I do not want to limit us in any way or define a closed group, having some working definitions and contexts for "wellbeing" here will help support our discussions. Several frameworks and indexes prove helpful as we deepen dialogs about different facets associated with the topic.

A Note About Trauma-informed Practices

In this book, I recognize the profound impact that trauma can have on an individual's wellbeing and interactions with digital environments. I am committed to supporting all readers who may have experienced trauma by adopting a trauma-informed approach throughout the content. This means that I strive to ensure safety, trustworthiness, and empowerment, and I deliberately aim to avoid any content that might inadvertently re-traumatize a reader.

I would like to invite everyone reading this book to explore this text in the way that feels most appropriate for them. I understand that trauma can affect individuals in diverse ways, influencing their emotional, psychological, and physical health. My goal is to provide information that is respectful, inclusive, and considerate of all readers, particularly those who have experienced trauma. I encourage readers to engage with the content at their own pace and comfort level. Everyone who is receiving clinical support is encouraged to seek out the advice and guidance of their therapist while encountering the examples, stories, and practices in these chapters.

Please be advised that while I aim to be sensitive to trauma-related issues, this book does not replace professional therapy or trauma-informed counseling. Also, if you find any content distressing, I recommend seeking support from qualified professionals who can provide tailored care and assistance.

I am dedicated to promoting wellbeing that is adaptive and personalized, and I hope that the approaches in this book help readers navigate the complexities of digital wellbeing in a way that is healing and empowering.

To Hyphen or Not to Hyphen: The Grammar of Wellbeing

As we dive deeper into the meaning of wellbeing, let us address the grammar and get that elephant out of the way: to hyphen or not to hyphen? Wellbeing is sometimes written as "well-being"—yet I choose not to hyphen it for this book because the hyphen is optional and one more unnecessary keystroke for us, and if you have not already figured it out, I'm all about poetry, efficiency, and flow. According to Google Ngram Viewer, which tends to focus on scholarly scientific and academic journals, well-being is about twice as popular as wellbeing, yet the preference is changing. I'm noticing a trend toward discarding the hyphen. Regardless, we are choosing unhyphenated for fluidity. Aesthetics are something, after all.

A Multifaceted Concept

How does the broader world define wellbeing? It's complex, but that does not mean it needs to be confusing. Most of us can agree that wellbeing is a multifaceted concept that encompasses various aspects of an individual's or society's state of being. It's often defined as a combination of feeling good and functioning well, which includes the experience of positive emotions such as happiness, and the ability to contribute to one's community. It goes beyond the absence of mental ill health, and it's linked to success at professional, personal, and interpersonal levels.[1] Many of us, including myself, are questioning that word "success" and redefining its metrics to focus on quality of life and a full sense of thriving, which is different than the measures of the past that tended to be purely economic.

Wellbeing can be described as "a state of positive feelings" and "a sense of meeting one's full potential in the world, which can be measured both subjectively and objectively."[2] It is a complex combination of a person's physical, mental, emotional, and social health factors and is strongly linked to happiness and life satisfaction.[3]

Wellbeing is a multidimensional construct that encompasses both hedonic (pleasure-oriented) and eudaimonic (meaning and self-realization) aspects. It is not limited to a single measure such as income, life satisfaction, or happiness, but rather, it is assessed using a variety of measures that capture these different dimensions.

Some people wonder about the differences, as we address terms, between wellbeing and wellness. Wellness is a term that medical professionals, including doctors, nurses, therapists, and counselors, traditionally use that tends to have a more clinical, health-oriented interpretation, whereas wellbeing encompasses many different disciplines and is often associated with lifestyle and the embodiment of wellness.

That said, this is also changing and more wellness professionals are adopting integrated approaches and addressing wellbeing as part of their methodology.[4]

The National Wellness Institute's Six Dimensions of Wellness gives a solid framework for what this interdisciplinary approach looks like. The graphic they publish (Figure 1.1) delineates six dimensions of wellness and maps out a holistic model that can be interpreted through a wellbeing lens. I like it because it shows interdependence.

Wellness is defined by the National Wellness Institute as a conscious, self-directed process that encompasses lifestyle, mental, spiritual wellbeing, and the environment, aiming for holistic, multicultural dimensions that contribute to a long and healthy life. Halbert L. Dunn, known as the "father of the wellness movement," introduced the concept of "high-level wellness," which emphasizes ongoing personal growth and change, where an individual is always "climbing toward a higher potential of functioning." The idea of wellness being optimal functioning within one's current environment is a definition that has been foundational in shaping modern

Figure 1.1 National Wellness Institute's six dimensions of wellness.

Source: National Wellness Institute, "Six Dimensions of Wellness," 2023, https://cdn.ymaws.com/members.nationalwellness.org/resource/resmgr/tools2/6dimensionssummary.pdf.

wellness perspectives. I respond to it and relate it to digital wellbeing, where we have talked about imagination, wonder, and awe helping us reach that higher, optimal plane of functioning.

I appreciate these definitions and interpretations because they are specific without being limiting. Wellness has been described as a state or quality of optimal health, and these dimensions deepen understanding and also address wellbeing from a holistic view without being vague.

Wellbeing Indexes Frame a Multidisciplinary Approach

Several widely recognized reports and indexes have uncovered sobering statistics. These trends could reflect those of the teams in your organization too: the Global Wellbeing Index (see Figure 1.2), which addresses Quality of Life, People and Community, and Education and Employment, surveyed 17,000 people worldwide and found that only 32% felt they had a good quality of wellbeing.

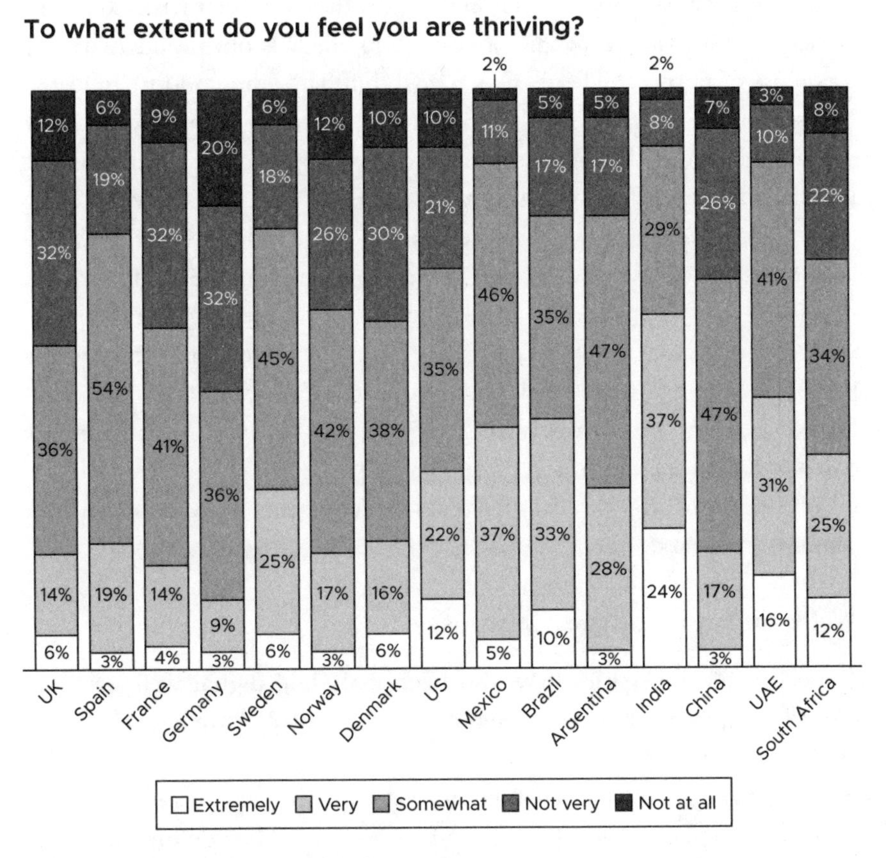

Figure 1.2 Global Wellbeing Index.

Source: Adapted from Longitude from the Financial Times and the Wellbeing Index.

Other prominent indexes include the Organisation for Economic Co-operation and Development (OECD) Better Life Index, the World Happiness Report, the UN Human Development Index (HDI), and the Gallup Global Wellbeing metrics, among others. All of these measures and indexes provide comprehensive views of wellbeing, capturing its various dimensions and providing valuable insights into the state of wellbeing at individual, community, and national levels.

It's my prediction that with more technological advances, namely those in areas of machine learning (i.e. artificial intelligence), wellbeing, creativity, and wonder will rise even more as a priority and imperative, not just in a one-on-one basis but across nations and political and ideological divisions. Why? For so long, our human progress was measured by "productivity" and ways to increase it. This served to make us adept at taking on more roles, aiming to multitask, and leading busier lives that increased toxic stress and anxiety. Did it make us happier and more fulfilled? Is our productivity tied to a sense of purpose? There's now a tidal shift happening, and it's universal. The sea change is that the rise in computing power can ultimately automate a lot of repetitive tasks and ease workload, if we approach it with reason and insight and design effective integrations. A human's role, now, can be to effectively look inward at how we address the human condition itself, as well as look outward at how we treat each other in our communities and in society.

There is an opportunity in this time to look intently and deeply at wellbeing and place it in priority, in all of its multifaceted contexts. If we no longer need to optimize productivity to lead happier, healthier, more fulfilled lives, what are we optimizing for? I would say it's the thriving itself that takes priority, and that's through wellbeing that is wonder-rich and creatively inspiring. Now, it's up to us to define the "how," and we have strategies to help.

Wellbeing on College Campuses

Colleges and universities have also been including digital wellness as an important dimension into varying models that usually have between 6 and 10 facets.

For example, Northern Kentucky University has a model where "Digital" is one of their 10 dimensions of wellness. They define the digital component as: "Digital Wellness is setting healthy boundaries and limits around your use of technology and screen time. A digitally well person considers the impact of virtual presence and use of technology on their overall wellbeing by taking steps to create sustainable habits that support their values, goals, community, and safety."[5]

While I find this definition useful in supporting healthy habit-building, I find it lacks the fullness of life design, and it addresses what to avoid in the language much more than discussing applications that extend creativity and increase positive connection. How can our approach to wellbeing, especially digital wellbeing, be more holistic and imagination-rich? How can those healthy boundaries be framed as intention-setting and connected to larger values and visions?

Digital Wellbeing Through a New Lens

So, how exactly could digital wellbeing play out, using our new lens of wonder and imagination?

Let us drill down, because what is in the media is not always the best approach to define this term.

Some articles, such as *Psychology Today*'s "What Is Digital Wellbeing," will attempt to boil down digital wellbeing as "the extent to which our digital lives help or hurt our wellbeing."[6] This is not exactly clear because the notion of a "digital life" is one that we cannot just take at surface level. It's a deep concept with a lot of nuance.

Some people would assume that digital wellbeing involves recognizing the power of technology and setting limits on our digital lives. This is a headline, as we see events posted for digital detoxes and tech sabbaticals.

But that's not the whole story. As the article points out, ". . . setting limits can indeed be helpful for wellbeing. But if a digital wellbeing tool's primary purpose is to help us be on our phone less, this means that it has an inherent assumption that more digital interactions lead to worse wellbeing. And the research does not support this assumption."[7]

A Dynamic Constructㅤ

I prefer to look at digital wellbeing as a dynamic construct. It's highly personal and context-driven.

I prefer to look at digital wellbeing as a dynamic construct. It's highly personal and context-driven. What one person finds beneficial to their wellbeing might not work for others. There is no prescription for relationships that are complex and evolving by nature.

Also, and most importantly, our expansive interpretation of digital wellbeing needs to take into consideration the incredible 3D technology that has a power to expand and extend our minds and ways of connecting and being with ourselves and each other as embodied humans. This is why we are exploring topics such as virtual reality and spatial computing, because of all of the imagination, awe, and wonder-rich opportunities to bring us and our teams into a better state of being, increasing our wellbeing.

Mariek M.P. Vanden Abeele, a wellbeing researcher, points out the modern limitations with existing approaches to wellbeing models. She presents a new and compelling holistic solution: "I argue that these constellations (of person-, device- and context-specific factors) represent pathways to digital wellbeing that—when repeated—affect wellbeing outcomes, and that the effectiveness of digital wellbeing interventions depends on their disruptive impact on these pathways."[8]

It's profound research with many implications. Vanden Abeele brings up phrases such as "ubiquitous connectivity" and "technological unconsciousness," terms that have to do with the ever-present expectation that we are all always "online" and connected, thereby taking it for granted and weaving it into our cultural and behavioral expectations for social engagement, work life, and all else.

The choice of how we go about weaving that connection is the ultimate consideration, and by introducing context and constellations as dynamic influences, we can get to the heart of the matter that is often oversimplified by mainstream media: there is no easy formula for digital wellbeing and no simple set of rules to follow. On top of this, technology itself is not a bad thing. It is not dividing us. Yes, it is ever-present, and there are addictive components to consider. Vanden Abeele shares, "Studies show that we hardly disconnect. Smartphones are tapped, swiped, and clicked over 2,600 times per day, and people spend close to three hours per day on their little screens—a figure that easily goes up to five hours and more for heavy users." These statistics were from 2019 to 2020. Imagine the numbers now, and we have spatial computing on the rise. It's no longer only screens that are drawing us in with their allure. It's worlds.

Digital wellbeing is a multifaceted concept that pertains to the healthy and conscious use of technology. It's about how individuals engage with technology and use it to enhance their lives, fitting into the broader spectrum of general wellbeing. Digital wellbeing interconnects with all other areas of overall wellness. It's universal.

Think about the teams you are a part of and serve. Are some of you working on different schedules and in varied locations? How has your distributed team been able to keep a sense of connection as you complete projects? Where, when, and how are you able to access the most creative inspiration, get deep work done, and feel a sense of wonder and awe, fueling engagement and passion? Do certain digital tools help you to collaborate and accomplish tasks? Are others more distracting, as they interrupt your focus and flow during the day? How are you able to show solidarity and respond as needed to those on your team, using a combination of skills, including emotional and cognitive domains? This book looks through those lenses and takes an approach that is both personal and professional, as we address our humanity in this tech-mediated landscape. I invite you, as you read, to think about the context of your professional role and the teams you are part of, and use the questions at the end of each chapter to frame reflections about how the topics throughout this book can impact and transform your whole life, personally and professionally.

Using technology in a way that promotes physical and mental health involves designing creative experiences with technology that encourage healthy use and mitigate negative effects such as overuse, dependency, anxiety, and stress. Digital wellbeing involves having a healthy approach to the digital world and all its opportunities. We can adopt a balanced and sustainable view toward digital media, ensuring that we enjoy accessing media and have a positive digital experience that involves agency and choice rather than compulsion.

I recommend focusing on the quality of wonder, and the relationships that can be cultivated using technology as a bridge and a medium instead of a divider or, ironically, a *screen*. It should not divide us from ourselves and each other. It's about the conscious and healthy use of technology, balancing its benefits and drawbacks, and integrating it into daily life in a way that enhances overall wellbeing. We need to talk more about approaches of mindful media—technology should be a medium facilitating connection in all forms. This book is part of the conversation.

Global Indexes Measuring Digital Wellbeing

Many global indexes and measures address digital wellbeing and integrate it with overall wellbeing, including the Digital Quality of Life Index (DQL), which examines digital wellbeing across 121 countries, representing 92% of the global population.

Another measure is the Digital Wellbeing Index (DWBI), assessing Gen Z's online psychological wellbeing. It uses the PERNA model, a variation on an existing wellbeing theory, comprising 20 sentiment statements across five categories. The first DWBI stands at 62 on a scale of 0–100.[9]

Other indexes include the Global Digital Health Index (GDHI), a global interactive web-based resource, and OECD's Digital Wellbeing Measures that uses a standardized module in information and communication technology usage surveys.[10]

These indexes and measures integrate digital wellbeing with overall wellbeing by considering various aspects of life impacted by digital technology, such as internet affordability and quality, overall mental and physical wellbeing, and electronic government services. They recognize that digital wellbeing is an integral part of overall wellbeing in today's digital age, and they aim to promote healthy and conscious use of technology.

As we look at a range of approaches to defining wellbeing, adding the layer of digital that is essential and linked to our overall wellbeing, we are reflecting an integrative mindset that can help guide each of us about how to live, work, and perform at our best. By viewing digital wellbeing as a dynamic construct, our conversations and curiosity-driven investigations in this book take on more dimension, complexity, and nuance.

We're bombarded with noise in all directions in so many of the spaces we spend time in. There are so many competing entities vying for our attention, most especially the one we so often carry in our pocket. Thus, much of the foundation of digital wellbeing can be framed in how we pay attention.

How to Pay Attention

Time to put your device away. Sound familiar? We hear this all the time. *Put your phone in airplane mode*—and now our planes have Wi-Fi so we quickly change our settings and jump back online. We're asked to use silent mode for concerts and theater, focus mode for meetings uninterrupted. No flash in the museum, no photos in the private club, no recording at the event. Our devices can do it all, and we manage keeping ourselves in-line, online, and very occasionally offline, which proves harder and harder by the year. We're slipping.

I held a meditation experience recently in virtual reality (VR) in a vivid setting of the aurora borealis. It was an hour-long event, a shared live experience in a social space. It was opt-in. I asked every participant right at the beginning to put away their phones, to keep them on silent. This was an in-headset gathering in a social, shared, spatialized platform where each person embodies an avatar in a beautiful virtual environment. The experience had movement, music, and meditation. As of the date of this writing, to my knowledge, there is no way to accept live phone calls in live online VR spaces. That said, I knew to announce the housekeeping rules at the beginning because people can still text at the same time, reading through the "nosegap" of a headset. I've been guilty of it myself—it makes you painfully cross-eyed, but it's possible!

When I told my guests that we would be celebrating an hour of *full presence away from devices*, they chuckled and nodded their avatar heads. Opt-in was universal. Later on, in the final meditation, I was in the middle of sharing a poem. Everyone was deep in focus, in collective flow, when all at once, in this landscape of snowy mountains, purple sky, and glowing holiday lights, the loud sound of a phone call—in the most iconic ringtone imaginable—reverberated on everyone's headset. Someone had left their ringer on loudly on their cellphone nearby and also forgotten to mute themselves in headset. Every participant in this meditation had to listen to the familiar e-ringtone cutting through the peace. Though I could have made a joke, I said nothing and kept the meditation fluid . . . in a few seconds, the jingle was gone, and the poetry and peace persisted. Still, the message was clear to me: even for dedicated meditators, this is a hard proposition. It's hard to be "away" for any length of time, and we create all kinds of excuses to support our constantly online, always reachable behavior. And at what cost?

During the period of writing this brief story, I've been tempted to check my messages at least 20 times. I convince myself that it will serve as a flywheel, boosting my energy and letting me come back to the writing with even more focus. Notes from friends are a dopamine boost, after all. It's a positive feeling, an inspiration, a surprise. Who knows what will pop up, or who will respond to something we have put out into the world? Lately, it's not enough to "like" something with hearts and thumbs up; it's also about the engagement of comments and messages. Content is still queen, but the algorithms are more complex. The unpredictability of the platforms add to their addictive properties. Even for me, knowing all of this, I still have to set boundaries, choosing to participate in certain ways that seem to me to be beneficial. I do not know if I get it all right. I'm sure I do not. I'm trying, and the conversation is an important one. There's a lot at stake.

That said, *have you put your device away yet*? That wasn't a joke, because I could picture you reading this book with your phone at your side. Can you put it somewhere farther away than arm's reach? Can you set it on silent? Do not disturb? If you are reading on a computer or tablet, iPad, or phone, could this be a solo exercise, the only thing you have "open" right now? Let us be soloists. Let us engage tête-à-tête, with a solitary focus on this voice, this exchange. I'd like to see what happens. Are you curious? I am.

I'm guessing that the first thing that will happen is you'll become a bit nervous. You might feel as if time is slowing down . . . you are starting to practice the mindful art of *reading* again, in conversation with the words. Your body and mind—note that this is a natural subconscious act, not a deliberate one—are enjoying the pace of digesting material word by word, sentence by sentence. We humans are meant for these types of encounters, ones where we can confront words, understand the meaning related to material that came before, then predict what will come next, then discover what is revealed, and then digest it and make sense of it and provide our own layer of meaning to it.

We use our imaginations in these readings, and we provide context and significance. That is what our brains are doing. We are not just consuming information. We're interacting with it in a relational way, an emotional way, and in a way that is additive in that we produce new ideas and understandings.

This is how I invite you to read this book. In this fluid, open way—not interrupted by phones and devices, not in a way that feels forced and unnatural. The reading part will actually happen naturally if you create the best conditions for it. Ironically, we are living in a time that has created challenging conditions for us to do this, for all sorts of reasons. The reasons do not justify it, though.

Tristan Harris, who cofounded the Center for Humane Technology, uses the example of a chair when he talks about goals for humane technology, which we do not have right now because of essentially a design problem. Our technologies were supposed to empower creativity and freedom and take us places where we can be agents of positive change, able to live lives of even better flourishing. That has not happened in this current period of tech advancement, and we are not charting a course for that right now. One of the major issues is that the technology system we live in is not ergonomic for humans. Take that example of a chair.

When chairs are ergonomically designed, they fit us—it's not us conforming to the uncomfortable chair (though that's the case many times in our daily lives!). Harris says, "If the chair is ergonomic (a metaphor for humane tech), it would be resting nicely against my back, and it would be aligned with the musculature of how I work. A chair that's aligned with that does not give you a backache after you sit in it for an hour. And I think that the chair that social media has put humanity in is giving us an information backache, a democracy backache, a mental health backache, an addiction backache, a sexualization of young girls backache. It is not ergonomically designed with what makes for a healthy society. It can be. It would be radically different, especially from the business models that are currently driving it." A great deal depends on redesigning that chair, especially with artificial intelligence (AI) in mind. It depends on our decisions right now. More on that in the chapters to follow.

Digital Wellbeing as a Part of Mindfulness, Wonder, and Awe

The topic of digital wellbeing is a burgeoning field, with active research and case studies impacting the way that we design and integrate technology into our daily life. It's a growing area for consideration, given that after encountering and undergoing a global pandemic we do not have clear boundaries between "work" and "home." Perhaps we never completely did, and we most likely never will again. So, the notion of a "work/life" balance will be modified in our conversations to reflect a holistic life design strategy that incorporates technology use in a mindful way. The related talks, research, stories, and use cases investigate the most current related issues as we aim to uncover best practices as well as emergent technology innovations in areas of wellbeing, incorporating our own life experiences in reflection. I'll raise questions for us to consider that I've been weighing myself on a daily—if not hourly!—basis.

I remind myself of my own ability to choose how I approach technology. Remember that we have a choice about the quality of our hours and days here on earth. It's the most valuable thing that we have. Time. And how we experience that time depends upon our quality of attention and where we are placing it. It's the *quality* of each element that we will be talking about a lot, applying research, concepts, and stories to our investigations. This is a book for us.

In the Introduction, we engaged in an exercise in mindfulness, used to open up wonder and imagination. I describe mindfulness using **the Three A's: aware, advancing, authentic**. I use those terms to make an abstract subject and buzzword more approachable. Over the years, those Three A's have served as anchors for a practice that can take many forms, from running a work meeting to composing a poem, from going on a hike to experiencing a collaborative environment in virtual reality.

When I first began to practice mindfulness, I did not necessarily call it that. I just knew I was directing my attention in a particular way, tuning into the present moment, and connecting it to my sensory experience in that time, without judgment, analysis, or interpretation. I was moment-by-moment experiencing a vitality of life in what felt like a pure flow, nonattached, able to access the rawness of the experience and disappear down a spiral into the very heart of it. I felt this is what Emerson was doing when he talked about being a "transparent eyeball," witnessing something in an act that was integral to presence. Losing himself.

I did not connect that sort of simple phenomenon to anything spiritual. I simply knew it felt special. I did not worship this act or lift it up in any sense . . . I just naturally practiced it anywhere. This was when I was fairly young. When I was doing something we might call mindfulness, it could look a lot like anything from deep focus to daydreaming, depending on what I was attending to. On a drive, staring out the window at the raindrops traveling sideways across the window, this was mindfulness. Looking at clouds? Mindfulness. Practicing lines for a theater production or learning about how a scientific principle affected the nature in my backyard? Also mindfulness. I was paying attention on purpose, creating context, and losing my "self" in the wonder. In my 2020 book, *Designing Wonder,* I connect these accessible, deliberate acts to the facets of wonder themselves, a meaningful part of integrated learning. The practice of mindfulness helped bring me into an active, conscious relationship with the world, one of belonging rather than merely passing through it mindlessly.

Have you ever taken time to attend to something that seems ordinary, finding the extraordinary in that simple thing? I call this "finding extraordinary in the ordinary," and I've given talks about that simple act of mindfulness and how it can be a gateway to animated moments of everyday awe. Tim Brown, an IDEO legend and renowned creative voice, famously recommended, "Once a day, deeply observe the ordinary." It can be anything that captures our attention, animate or not. It causes us to see with fresh eyes and then appreciate everything anew.

Finding the Extraordinary in the Ordinary: Mindful Awareness and Wonder

There's a beautiful scene in the 1993 film *Blue* by Krzysztof Kieślowski, one of the most iconic and influential film directors of all time. Juliet Binoche holds a sugar cube carefully over a cup of coffee, delicately dipping one small corner in, attending to it. The coffee saturates the cube slowly, and we watch it. Then, the cube is dropped into the liquid with a satisfying "plop." This scene took many iterations to film, with care. Kieślowski put it there as a mindful interlude, also showing that the character attends to her surroundings and has patience with these things. She is able to dissolve into this quiet experience of slowness, of noticing. He has given a masterclass on the scene itself, showing the intention that went into creating it and his personal philosophy underneath it.[11]

This also describes what mindfulness can mean: *attention coupled with intention*. Directing our attention, our most precious resource, with intention. Using intention to inform how we manage our attention. It's all about mindfulness, and it's something we have as humans. So far at least, this quality of mindfulness does not apply to AI. Mindfulness is highly infused with personal emotion, even as we also acknowledge the universality of what we share, detaching from forced control over outcomes.

I've come to believe Kieślowski was a master of showing this sort of mindfulness and wonder, and had a way of making the internal noticing experience visible on film, which then translated emotionally to the audience. He was mindful of the complexity and pervasiveness of emotion and said of his work in an interview at Oxford in 1995, "It comes from a deep-rooted conviction that if there is anything worthwhile doing for the sake of culture, then it is touching on subject matters and situations which link people. . . . Feelings are what link people together. . . ." He had a way of making the inner workings of the mind and heart visible and relatable in a visceral way.

Engaging with Mindfulness, Wonder, and Digital Wellbeing

I was listening to Ellen Langer in an interview she gave about the power of being mindful instead of mindless in a digital world. Langer offers fresh candor and a positive take on the subject that is backed by scientific rigor. Her sense of humor is infectious, and her logic emotionally compelling, rationally convincing, and invigorated by a career of helping others live lives unshackled by autopilot.

A renowned Harvard psychology professor and pioneer in the field of mindfulness, Langer has extensively explored the impact of technology on wellbeing through research, public discussions, and a fair share of debates. Her perspective emphasizes the potential for technology to foster mindfulness rather than detract from it, challenging common perceptions that technology inherently contributes to stress and disconnection.

I agree and choose to think proactively about how we can be more intentional about our technology use rather than vilifying it, and consequently shaming ourselves, which does nothing to improve our habits and overall outlook and self-compassion. Langer similarly argues that the issue is not technology itself but *how we engage with it*. She suggests that technology can be a tool for enhancing creativity and wellbeing if used intentionally and mindfully. For instance, technology offers endless opportunities for creative expression and can facilitate meaningful connections with others, which can improve the quality of our lives and relationships.[12] We can all become more aware of our reactions to technology, thereby mitigating stress and enhancing our emotional stability.[13]

The issue is not technology itself but how we engage with it.

Langer provides a compelling example of how leaders who are open to adopting new technologies have better success: "When new technology for cardiac surgery was introduced, those medical teams with leaders who minimized concern for status differences—in other words, were willing implicitly to admit they did not have all the answers and take advice from underlings—had the most effective communication, learned the most, and found the transition the easiest."[14] Her belief in the value of humility and openness to learning, even in highly technical and specialized fields, aligns with growth mindset and what science has proven to be most advantageous.

We can all apply this to how we approach our roles on teams and our communication with others in relationships. We can be curious instead of afraid, humble and open instead of fixed and overly dominant. We do not need to fear new technology; we need to be mindful of larger goals and how it can serve our thriving. When we are mindful of something, we imbue it with meaning and intention. Langer's work emphasizes the role of mindfulness in transforming our interaction with technology.

We can use technology in ways that promote creativity, foster meaningful connections, and support our wellbeing rather than allow tech to exacerbate stress and disconnection. By openly and curiously reevaluating our relationship with technology, and infusing wonder and awe into the equation of looking at how we can use tech to connect with ourselves, each other, and nature, we can better harness its benefits to enhance our wellbeing and imagination.

The Power of Transformation Through Mindfulness and Wonder

How can we each apply such open mindsets on personal and professional levels, in a way that meets us where we are at and allows for transformation? That is the essence of what we explore in this book: ways to rise to that higher plane. It's one of the explicit goals of this book, to provide those inroads and avenues with new methods.

When we practice actively noticing new things around us in our environment, we place ourselves in the present moment and become even more sensitive to context and perspective.[15] That level of engagement with the present moment, and a heightened awareness of surroundings and experiences, can also make us better at attending to the needs of others. We become better listeners and responders, adaptive to what's emergent around us. Even just five minutes per day of a "noticing walk" is a way to practice tuning and grounding ourselves. This does not necessarily involve meditation or yoga, though it could. It's a simple secular act of noticing, open to anyone.

Research over the past four decades has demonstrated that these types of mindfulness and wonder practices can reduce stress, unlock creativity, enhance performance, improve health, and increase life satisfaction.[16] Stepping into awareness is the essence of engagement and is energy-begetting—not energy-consuming. It enables all of us as individuals to value ourselves and practice self-care and self-love, also extending that care and love to others simultaneously. We better recognize and take advantage of opportunities, avert risks, and improve interpersonal relationships by making them less evaluative and more charismatic.[17]

For years, I've cited Langer's groundbreaking studies as examples of how transformational open mindsets can be. One of the most notable, the "counterclockwise study," illustrated the power of mindfulness to affect physical health and perception of age. In this study, older participants who were placed in an environment that replicated the past and encouraged to live as if they were younger showed significant improvements in physical health, appearance, and cognitive abilities.[18] The study underscores the mind-body connection and the potential for mindfulness to influence both psychological and physiological wellbeing.

By focusing on active noticing, which engages individuals with the present moment and enhances their sensitivity to their environment, mindfulness has direct benefits for mental and physical health, imagination and creativity, and interpersonal relationships, offering a practical and accessible approach to cultivating mindfulness in everyday life.

Addressing our work lives, we can use these practices to challenge some of the limited conventional understandings of wellbeing, mindfulness, and creativity through wonder, by demonstrating practical applications. We can integrate these mindsets into daily life through active engagement with and observation of our surroundings. Later in the book, you will see how layers of immersive digital applications also add new facets that are expansive in inviting us and our teams to explore with a playful, curious embodiment of wonder and awe. These approaches can lead to more fulfilling and engaged lives that are rich in appreciation and awareness, inviting experiences to lead us there. We become more creative in our problem-solving, more adaptable to new situations, and more liberated in our notions of self and others, among many other benefits.[19]

There are great applications here for this type of wonder-infused digital wellbeing worldwide, especially when translated and applied across cultures that have different spiritual beliefs and contexts. It's approachable and scientifically sound. This becomes a complete gamechanger.

Psychologically Safe Workspaces and Their Impact on Imagination, Wonder, and Creativity

Adopting, defending, and cultivating a **psychologically safe workspace** makes such a difference for us and our teams as we look to expand mindsets and adopt curiosity, wonder, and awe into our practices. Some of us might be experiencing topics in this book that we want to implement with our colleagues and collaborators. How do we do this bravely and exuberantly, with an open sense of possibility? We started out as highly creative beings, and it's essential to defend that wonder and imagination that came so naturally to us as children.

I think about Sir Ken Robinson's TED Talk—the most watched TED Talk of all time—titled "Do Schools Kill Creativity?"[20] With nearly 77 million views worldwide, it's made a significant impact. It's profoundly tragic that so many of us lose our sense of creativity and wonder over time, and this loss has a lot to do with framing creativity and awe as psychologically safe topics in all environments—learning and professional environments included. How often do we bring wonder into our boardrooms? We need to frame it as a safe topic, a must-have rather than a nice-to-have, especially at a time when the World Economic Forum has cited creative thinking[21] as the number one "future of work" skill. It's nonnegotiable.

It's not easy to live a creative life and to keep wonder and awe alive in our minds and hearts. This book addresses that explicitly. We can use these practices, and psychologically safe spaces, to foster learning, innovation, and growth. Amy Edmonson, a professor of leadership and management at Harvard Business School, wrote a 2023 book, *Right Kind of Wrong: The Science of Failing Well,* to provide a framework for understanding and practicing failure wisely and adopting psychologically safe workplaces. Edmondson outlines the three archetypes of failure—basic, complex, and intelligent—and emphasizes that "intelligent failures" are essential for learning, innovation, and success. This is essentially all about mindfulness, wonder, and related qualities of being open. Current research is showing us that we can embrace human fallibility, replace shame and blame with curiosity and vulnerability, and develop a healthy relationship with failure to pursue smart risks and prevent avoidable harm for ourselves and the teams we serve.[22]

My background as an immersive technologist, programmer, poet, author, experience designer, educator, and interdisciplinary team leader, among other roles, involves both the secular and the spiritual, and I honor and celebrate the overlaps and integrations. I have gathered and created many wonder-and-awe rich, mindfulness-infused resources in this book, which can serve as your core principles and design flows, helping you to defend creativity and help foster psychological safety for your team.

As a leader, it's been one of my biggest takeaways: my teams achieved more success, individually and collaboratively, when they felt safe and encouraged to get messy, get creative, and to embrace kindness. To be human is a privilege, not a requisite for getting work done in this day and age. We need to learn to amplify our creative humanity and to access awe and wonder to reanimate our inner lives connected to the outer work, bringing us back to what is most essential and special. We each matter as part of our communities, and we are not meant to exist in isolation. That's one of the biggest lessons on diversity, inclusion, and collective resilience: we are better off feeling safe, encouraged, and connected to each other. The invitation is to show up as we are and to share, learn, and depend upon one another for support, collaboration, care, and inspiration. Life is meant to be explored, not performed.

Throughout this book, I'll bring in useful, practical, expansive creative frameworks and principles that I use as part of the core of my practices. I invite you to use them too. To find out more and continue your own development, even beyond the journey of this book, you can visit my website at caitlinkrause.com, where you can explore online resources and courses, and also find links to my additional publications including *Mindful by Design and Designing Wonder.*

Frameworks for Mindfulness and Wonder: Using the Three A's as Guideposts

I introduced mindfulness earlier in this chapter and explored how we can apply awareness, advancement, and authenticity to our life experience design. These Three A's are deeply explored in *Mindful by Design*, and there are many more resources about mindfulness practices linked to wonder and awe in *Designing Wonder.*

Looking now at approaches to teambuilding, collaboration, and creativity with a mindfulness lens of the Three A's, we can see how being more **aware** can lead to appreciation, how a mindset tuned for **advancement** and growth keeps us adaptive, and how **authenticity** encourages us to have relational trust and psychologically safe communities where we operate from curiosity rather than fear. So many links and extensions evolve from here.

An appreciation of these three qualities, attending to them, is closely linked to **wonder and awe**, which is what we'll discover more about in the following chapter.

Throughout this book, we will introduce stories and subjects that are all facets associated with the broad subject of digital wellbeing, a topic that is shaping the future. We will adopt a critical and curious lens and also explore how each issue affects us personally and professionally, for ourselves and the teams we serve, with reflection questions along the way. The "Reflections for Teams" are especially geared toward serving leaders, and they can also inform how you approach your relationships. I will also include my own "Extensions for Future Exploration" section at the end of each chapter, indicating where I feel drawn to expansions. If you would like to hear from me about the extension topics, let me know. A sequel could be in the future.

Personal Reflections

1. What is your best design strategy for increasing your attention and focus?
2. How does the analogy about ergonomic technology resonate with you? What intention do you have for your own digital wellbeing becoming more "ergonomic" and less uncomfortable, if it feels that way?
3. How is your overall level of ability to access states of wonder, awe, and creativity? When in the day do you find your creative attention lapsing, and when is it strongest?
4. How are you inspired to approach mindfulness, using the lens of the Three A's?
5. Do you have any practices currently that involve letting go of self-consciousness, having a certain mindful quality of noticing the extraordinary in the ordinary? If yes, describe how it feels. If not, describe how you might like to create time and space for this.

Reflections for Leaders

1. What has been your most effective strategy when it comes to planning your day and establishing healthy habits with technology for you and others?
2. What new strategies would you like to adopt that can increase your awareness of your surroundings and increase your ability to listen and be heard?
3. What have you encountered in this chapter that resonated with you or reached you in some new way?
4. How is a psychologically safe workspace that fosters creativity and risk-taking meaningful to you?
5. Have you ever had an experience with "intelligent failure"? If so, how did it feel? If not, how could you imagine embracing it?

Extensions for Further Exploration

Drawing from conversations in this chapter, extensions that intrigue me involve applying scenarios directly into immersive environments designed for imagination and wellbeing. Imagine, an "intelligent failure" sandbox designed explicitly to spark curiosity, improve team dynamics, and allow for creativity and collaboration to bloom. Mindset shifts are meant to happen here. There are so many ways to bring more dynamic qualities into application design, especially with wellbeing as a priority. We will explore many of these expansions in the chapters to come, too.

2 Frameworks for Digital Wellbeing

Through our new lens of wonder, digital wellbeing becomes a dynamic construct.

It's not a formula. Treating it as a formula or a prescription for behavior is dangerous because it leads to a toxic loop of self-assessment along with a binary view of good/bad behavior in response. Too often, we have inadvertently treated our relationships with technologies with a hidden or overt shame. We feel as if we are underperforming, and this is part of the destructive motivation that social media and gaming has used in the past to influence our behavior.

It's no surprise that, in the past, conversations about wellbeing and digital wellbeing have been more concerned with setting boundaries than exploring possibilities. We teach our children to do this too, to focus on limits over luminous boundaries instead of what's bountiful and abundant. Our focuses have been restrictive and fearful rather than expansive, awe-inspired, and joyful, and our creativity and heart-centeredness have suffered as a result, leading us toward isolation, loneliness, and mistrust.

Of course, safety and responsibility are critical and essential. This is a given. When we set the right foundations, we can grow a new system. Our current view of wellbeing integrated in a technology-rich world is static when it *can and should* be dynamic. Some would guide us to set limits and conditions onto our behavior patterns in efforts to govern ourselves and others into submission rather than learning about and applying what works well for full life flourishing and thriving.

A new generation showed us the power of gaming to inspire, to enlighten, and to share agency, and those in power responded (some of the time) by creating closed systems, preventing interoperability, and employing mercenary design techniques that engage to choke, addict, and disempower. Some of these systems, including the sad state of social media, left many of its users feeling depleted rather than uplifted.

It does not have to be this way.
It should not be this way for our future.
It's time to change that narrative and flip the script.

I believe in invitation over prescription when it comes to wellbeing, wonder, mindful media, immersive design, and gaming technology. It's possible to change our entire view toward these subjects without disrupting our day-to-day functions. It's a reframe with exponential effects. This chapter explores and animates several key frameworks as well as examples we can draw from to make healthy lifestyle choices that prioritize our sense of agency. Throughout this book, we explore applications in areas including gaming, spatial technology, learning design, music, movement, integrative biofeedback and more, and see how these arenas can support wonder-animating digital wellbeing.

We can have brighter experiences through this transformation of mindset and practices. Right now, right here, we have an opportunity with the rise of AI and computing power to focus on what makes humans thrive at their best. Technology, especially digital interactive technology, has the most power right now to lead that change we want to see. Why is that? Because it is motivating, it is fun, and it has the power to reshape our realities.

This book, ultimately, is about reshaping the narrative. It's about seeing the ripe possibility of digital wellbeing to transform lives and enhance them in ways that offer better solutions than we have had in the past.

This book, ultimately, is about rewriting the narrative. It's about seeing the ripe possibility of digital wellbeing to transform lives and enhance them in ways that offer better solutions than we have had in the past.

Right now, loneliness is a global crisis. Right now, huge swaths of our human population are living feeling disconnected and lost. At the same time, a digital revolution is underway that gives the opportunity for the whole planet to be connected like never before. Spatial computing represents one of the biggest shifts.

What Is Spatial Computing?

It's all over headlines and tech announcements, but what exactly is it? Spatial computing is a new form of computing that uses AI, computer vision, and extended reality (XR) technologies to seamlessly blend virtual content into the physical world, breaking free from traditional screens. We no longer have to communicate and connect using awkward 2D modes.

XR encompasses a continuum of technologies blending real and virtual environments through computer technology and wearables. VR sits at one end of this spectrum, offering completely immersive experiences through devices like VR headsets, such as exploring simulated underwater worlds. Moving along the spectrum, augmented reality (AR) overlays digital information on real-world views, typically through smartphones or AR glasses, enhancing physical sites with digital data. Between these is mixed reality (MR), which integrates real and virtual worlds, allowing physical and digital objects to coexist and interact in real time, such as in educational tools for medical students to interact with 3D models of human organs. This continuum from VR to AR, with MR positioned centrally, demonstrates the range of immersive and interactive experiences possible within XR. We will explore digital wellbeing in the context of XR more deeply in Chapters 7 and 8.

The core idea of spatial computing is to make computers and digital information more integrated with the physical environment, allowing humans to interact with technology in more natural and intuitive ways. It enables humans, devices, computers, robots, and virtual beings to navigate and operate in 3D physical spaces.

Spatial computing involves capturing 3D data about the physical environment using sensors like cameras, lidar, and radar, analyzing that data with computer vision and AI, and then overlaying digital content and interfaces onto the real world. This allows for new types of human-computer interaction beyond the traditional keyboard and mouse.

Think about all the possible applications that can make our lives easier, and all of the potential downsides that could increase the noise factor and the friction. Spatial represents a seismic shift in human-centered computer interaction, allowing for more contextual experiences, which we will investigate in later chapters. The point is that all spatialization is media too, and it's going to change the nature of what it means to be a human consuming and creating media in a tech-saturated world. Digital wellbeing is one of the most important considerations for humanity at this stage. We will soon all be coexisting in 3D spaces, alongside robots and virtual entities. We are already encountering this coexistence, and it's a globalized system that is growing in its applications and potential benefits.

With spatial computing on the rise, and new ways that accessibility and agency are increasing, it's time for media to focus on the stories, experiences, and exchanges that matter. After all, that is what "media" means: *between*. It's an exchange, a conduit. As media focuses on that meaningful form of connection, science can play a larger role in guiding us in what works best for the wellbeing of the human brain, mind, and body.

In this book, I'm inviting all of us to step into new ways of examining digital wellbeing, to uncover creative applications and learn how people are approaching and applying it in positive ways. We can use a lens of curiosity to approach our own lives. We'll explore and experiment with digital wellbeing, connecting it with wonder, awe, and overall life thriving. As mindfulness practices encourage, we will use our own bodies and our sensibilities as indicators, trying out ideas and seeing how they land with us. We are each our own best experimental, experiential testing ground. Each of us will have our own responses and reflections related to this material, and that's as it should be.

I'm encouraging us to get courageous with this material here, and to go beyond any labels we might be carrying about our personal and professional roles. To engage fully, our goal is to both appreciate our own multifaceted forms of identities—as Whitman has said, "We contain multitudes"—and to dynamically connect our inner identity and outer life, our work identity and the one we have at home, so that we can live what Parker Palmer has called "an undivided life." When our world can celebrate its own dynamic multifacetedness and also its wholeness, we experience a new sense of freedom and belonging, and we can bring that awareness, appreciation, and sensibility to our teams, personal relationships, and collaborators.

Regardless of age and stage, we should all have access to the wholeness and belonging digital wellbeing offers. It does not matter whether we grew up self-identifying as a "gamer" or not, whether wellbeing and wellness in general are something we are familiar with, whether we have adopted meditation and mindfulness practices, the list goes on and on . . . we can quickly label ourselves and others, to our mutual destruction. It's time for new ways where everyone can access new methods, make them their own, and discover ways to thrive.

This book defies labels and also aims to demystify terms and techniques. It's meant to be accessible and also integrative in its approaches. There is more curiosity and less agenda; I'm not here to prove anything. I'm here to question and find out alongside you.

The course I teach at Stanford University, called *Digital Wellbeing by Design: Healthy Relationships with Technology*, is an elective, and students will join from all different disciplines and backgrounds, with many different goals and aims that attracted them to the course. I added the "design" part to the title because we are all designers.

You can play a role, right now, in reshaping your life. Let's give it a go.

Foundations for Wellbeing: A Framework for Creativity Through Culture

Four Culture Cornerstones

In my 2019 book, *Mindful by Design*, I identified four primary components that serve as the foundation of mindful learning environments: Dignity, Freedom, Invention, and Agency. I called them elements of culture, or "culture cornerstones."[1]

These four anchors guide design strategies across digital applications, including spatial computing, gaming, and mindful media applications. As we explore digital wellbeing, I'm calling them the "Four Culture Cornerstones of Mindful Media" because everything is about transmission.

Dignity, Freedom, Invention, and Agency each have a ripple effect that amplify wellbeing in applications. We can use these four core tenets as a guide for how we lead and promote organizational change.

(*continued*)

(*continued*)

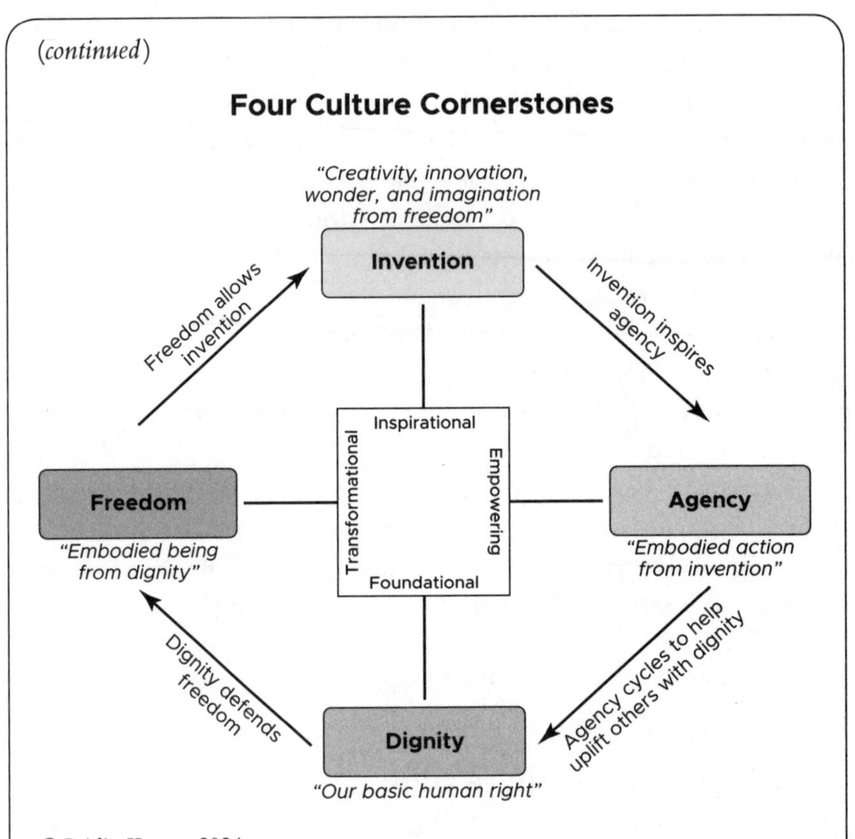

©Caitlin Krause, 2024

Why? Because they are essential, and they uplift the imagination process, thereby combatting stress, anxiety, and depression. When you address culture first, worlds open. (Note that it is impossible to address culture and uplift a sense of wonder and possibility without safety. We will see this play out in the process of the four tenets.)

I knew that all quality experiences are built around these four key tenets, and that they interplay. They can help define a culture and allow people within it to "come alive" in a sense—to grow and thrive and flourish.

We want to ignite imagination and wonder, to spark the creative impulse and then build something that lasts, something that transfers from individual to individual, spreading out to the world. Connection.

Let us talk about each one in more detail, thinking about how we could invite this awareness and intention into dynamics with teams we serve.

Dignity, in sum, encompasses a mutuality of respect. As I describe in *Mindful by Design*, "It involves actionable, palpable respect and care for each individual in the group. The art of listening, holding spaces, sharing with a mindful reverence for the 'other' as a source of insight, building awareness of connections rather than separations, embodying and acting out of respect for self and others." Honoring identity and core beliefs. Dignity is about belief, vision, and passionate integrity. It is about conviction of values and holding space for complexity and emergent stories that reflect who we are. Our voices can come alive from a source of groundedness, of sovereignty, of dignity.

Freedom is about really seeing each other, and asking as a collective, *how do I choose to be for you to be free?* People experience freedom when they are safe, both physically and psychologically. "Freedom, when operating from a place of dignity, means that each individual has a voice . . . It's a freedom to create new connections and form new understandings and inferences as well." It involves a deep self-awareness and social awareness where everyone feels endorsed and welcomed from the beginning. There's true collaboration that exists when open sharing is curiosity-driven and emergent, not motivated by something to prove or defend. There's a mutuality of freedom. How do I have to be for you to be free? It's about us. To be free is to be open to what is emergent in the moment, uninhibited, and to then invite that level of engagement to spark imagination.

Invention is about an environment of open wonder and creativity, where anything is possible. We've set the conditions, through dignity and freedom, for invention to loft. We can take risks, safely, because conditions are already in place to support these bold new ideas. "It comes from the *'What if . . . ?'* questions and opportunities to design possibilities. The cornerstone of Invention is about mindful problem-solving, creativity, and meeting real-world challenges with feelings of optimism, inclusion, compassion, and commitment." As artists, creatives, and inventors, we carry amazing insights and dreams,

(*continued*)

(continued)

and they are unbridled, fomenting something transformative. "The mindful inventions that come forward from here have the power to change the world."

Agency means that possibility and creative imagination take flight in empowered action. It's not enough to simply have dignity, freedom, and invention alone. These are all potentials, in a sense. We need to have a place to build and an empowered voice or means to do it. Agency is all about that. "Agency couples with identity, and transfers to mindful act." Depending on complexity, we can start as mentors by guiding and supporting individuals and groups in their actions. Eventually, they develop personal agency, and this is when relational trust meets evident action, reflecting the four cornerstones in consonance. Our actions are an iterative process, building toward actions that have positive and evident impacts.

To have a sense of agency is to belong, and to understand the connection to a larger system, with understanding and empathy for the value of each perspective and contribution. The beauty is, as social communities, we can and should all support each other in taking actions of empowered Agency, and in aligning it with larger visions and goals that are not about just one person. There's a strength in being together in communities of networked support.

If you think of any meaningful experience you have ever had, whether it was creating a product and shipping it, working as a team to develop an idea and bring it to life, or coming up with a new idea that needs to launch and loft—just think back and evaluate it from the perspective of those four cornerstones. Were they present? What difference did they make? If they were not there, how did that gap play out?

Media is a transfer. We want our great ideas to get from our own minds to a page in our private studio, and from the studio to the collaborative salon of our teams, and from that generative salon to the stage of audiences and public that can interact with them. Media is constantly interplaying in those worlds of *studio, salon, and stage.* These four culture cornerstones are value based and qualitative, framing how mindful media can develop well, with imagination, wonder and wellbeing in priority. They make all the difference.

Neural Networking for a New Roadmap of Open Pathways

For the past couple of decades, I've been using the brain's neural network as a simple model for showing how everything should operate in our digital learning and design strategies. This is no longer a world of centralized power when it comes to ideas. It's even moved from a decentralized network to a distributed network of rapid transfers of information deeply encoded in context and meaning. There are many nodes, and there's a simultaneity and an inherent beauty, not a sequence. In understanding the synchronicity, we need some mutual trust in our networks. In a sense, we are mindfully "neural networking" with each other all the time, even without consciously recognizing it.

Humans are social animals. In a distributed network that prioritizes creativity and agency, it's important to have trust and a certain "data diligence" and "data dignity" over the sovereignty of our identities and the data we choose to share. This all plays a part in how we can approach and look at the system.

Using the network of the brain as a metaphor can apply to so many areas of life. It's how the media has shifted, and this encompasses how our distributed networks of gaming and spatial computing are evolving, too. Everything becomes about *media*, or knowledge/experience transfer in a network. That is what media means, after all. A transfer mode. A channel. These can be open pathways. And it's not one direction. It's a matrix!

The reason that agency is of utmost importance here is because if you take away the agency, you take away empowerment. We do not want our media to do that. We've seen what happens when it works the other way.

As an experience designer and essentially a guide across a wide variety of the continuum, from digital spatial computing environments to physical ones, I can play the mentor and the support to enable all four culture cornerstones to coexist. We can all witness teams and cohorts take flight on a daily basis, and we continuously learn from those we lead, serve, and empower.

Now, when I lead a session for business C-suite executives, a module for TED, or a training for the Air Force, I use the Four Culture Cornerstones as the standards for my design strategies.

It's all based on experience. Of course, the content will vary and become very personalized. The values and architecture underpinning what we are doing and why we do it do not waver. These four tenets are critical, and they have many layers and interplay with other methodologies and complementary frameworks, many of which I will mention in this chapter and in future sections of this book.

These four tenets of Dignity, Freedom, Invention, and Agency can and should be applied to digital wellbeing modules. Whether building a game or a platform, an app or a tool or a device, all four should be anchors for the design. When you combine Dignity, Freedom, Invention, and Agency, a participant can thrive.

The four cornerstones provide a foundation for wellbeing, and they should be underneath all creative processes, and a way to approach mindful media design. All technology is media. It becomes more about the quality of imagination, and access to wonder-fueled pathways that connect us.

The cornerstones are simple, easy to remember, and one leads toward the next. The beauty, too, is that they are often self-guided and allow everyone to see where they are in a connected continuum. They start with dignity and build to freedom, invention, and agency. It's an active, interactive process, and the digital tool or experience becomes the *vehicle or form of media* for the person on the roadmap.

Experience Design Guided by Emotional Intelligence

I've recently made a compelling discovery: everything meaningful—everything that *moves* us—is embedded in *experience*. Our lives are a series of experiences, and emotions are how we encode experience. Designing experiences that support wonder-rich digital wellbeing must necessarily involve deeper awareness and understanding of emotional intelligence and mindful experience design. All of us are serving as impactful leaders, whether in business or beyond, as guides, parents, colleagues, friends–*humans!* This involves understanding how to connect well with others and engage in positive collaborative experiences so that we all can thrive.

Think back to times in your life when you have felt an emotion. In most if not all cases, something is *happening* that causes the emotion to occur.

We do not have any meaningful experiences without emotion, and our emotions hinge on the experiences we have.

We are emotion-rich humans who *feel things* by nature. This is a major part of what makes life worthwhile and absolutely astonishing for us: because of the emotions we feel. It is a privilege and a literal pleasure to be wired for emotion, unlike our AI-powered robot counterparts in this evolutionary time. Let us dive into more of what emotions mean for us as humans.

We empathize, we connect, and we use emotions as our motivation for embodied action. I find Lisa Feldman Barrett's research on emotion to be inspiring and astonishing in this arena, so before I get into the mechanics of gaming and the impact on immersive design for spatial computing (including augmented and virtual reality and the whole XR continuum!), let me first deconstruct some of Feldman Barrett's most mind-blowing recent findings.

Lisa Feldman Barrett is a distinguished neuroscientist and psychologist working at Northeastern University whose research has significantly contributed to the understanding of emotions, debunking long-held myths about the brain, including the triune brain theory. Feldman Barrett's work is characterized by its interdisciplinary approach, drawing from psychology, physiology, anthropology, philosophy, linguistics, evolutionary and developmental biology, computer science, engineering, and the history of science.[2]

One of Feldman Barrett's major contributions is her development of the theory of constructed emotion. This theory challenges the traditional view of emotions as innate, universal categories found in specific brain regions. Instead, Feldman Barrett proposes that emotions are constructed in the moment, through a complex interplay of several core systems, including interoception (the sense of the internal state of the body), a conceptual system (for the sense of the internal state of the body), a conceptual system for making sense of sensations, and the influence of the social environment.[3] This perspective suggests that what people commonly understand as distinct emotions (like anger, sadness, or fear) are not universally expressed or recognized but are culturally influenced and context dependent.[4]

Here are thoughts I've been crystallizing from Feldman Barrett's research, as seeds for future conversation. For me, they layer on top of Brené Brown's robust body of work and create a whole new framework for how and why we think and act the way we do as sentient creatures:

We, as humans, have a well-developed brain that is often guided by allostasis, meaning we are energy conscious. We have a finite amount of energy, and we are looking to regulate it and to maintain it at all times. We're also wired for our own safety, so perceived threats, whether physical or psychological, have a tremendous impact on our choices of action. Our brain is constantly predicting its world, and this leads, often, to us predicting ourselves into hallucinations rather than tapping into what reality actually is.

When others might look to judge our emotions by what is on our face, this is not dependably accurate. In fact, our facial expression is not a reliable indicator of emotion, as proven by scientific research. As Feldman Barrett points out, "Traditional views of emotion don't comport with evidence." We use phrases colloquially with others like *I feel emotion* or *I perceive emotion in other people*, but scientists currently cannot find a single biomarker that points to an emotional state. This is incredible!

Feldman Barrett's view, informed by extensive research, is that the brain is constructing instances out of basic ingredients. The brain is constantly making situational predictions, saying, "What in my past is similar to this situation?" It then constructs concepts, or categories, on the fly about emotions. One example Feldman Barrett uses is anger: "Anger is a population of variable instances in the brain that depend on what action you're about to take—they could be cry, punch, yell, etc." The physiology of these variable instances are linked to *action*. The body gives the brain some signals that are sensory, and they are not used for pleasure. They are used to inform the action.

The brain is preparing the internal system to move in some way, and it infers what you will sense based on those visceral motor predictions. "Your experience of the world and your experience of your own body are a consequence of those predictions, those concepts." From there, we apply concepts of emotions. "When your brain makes a concept for emotion, it is constructing an instance of that emotion. That's how emotions are made."

The brain's predictive mechanisms are operating so it can change the internal system to do something. Metabolically, we are then focused on execution, which is ultimately about moving, and also learning from the experience so that we transfer that knowledge to the next instance.

One of the most liberating mindsets I've adopted, given these breakthrough findings about emotion, is the understanding that we are each constructing our dynamic world in each moment, based on our past experiences. We are constantly reorganizing, restructuring, and remembering. I would say, given these conditions, that we actually have a *conscious choice* in our awareness and subsequent attachment of emotion, which will follow the action we take. As Feldman Barrett points out, emotions are constructed by the brain in the moment *after* the action, not before it!

We might be slow to recognize this because the slim fraction of a second that separates action from the assigning, experiencing, and encoding of emotion is imperceptible in our raw sensory awareness as participants in the action. After all, it's our story we are living, moment by moment—often without self-analysis in those moments! Also, it might be counterintuitive to many of us who were raised with a different understanding of emotions, which we thought lacked rationality and sometimes swept us into acts of passion. Learning that this is not the case can be entirely freeing and transformational.[5]

The further you get into the research, the more wonderful it unfolds, in my opinion, because it debunks all of those simplistic metaphors and maxims for the brain that might have sounded catchy but do not serve us well when it comes to truth and the complexity of our experience. Being complex does not have to mean confusing. One of the prime brain metaphors we need to discard is the triune brain theory. Proposed back in the 1960s, it suggests that the human brain can be divided into three distinct regions: the reptilian complex (basal ganglia), the paleomammalian complex (limbic system), and the neomammalian complex (neocortex). These regions were thought to be responsible for primal instincts, emotions, and rational thinking, respectively. However, modern neuroscience research has largely debunked this theory. It's now understood that the brain did not evolve in successive stages as the triune brain theory suggests and that the brain's functions are not as compartmentalized as the theory implies. Emotion and cognition, for instance, are interdependent and work together, contradicting the idea of them being controlled by separate brain regions. While the triune brain model is an oversimplification and not accurate in many respects, it continues to be used today due to its simplicity and ease of understanding.[6]

Feldman Barrett and other contemporary scientists argue that the triune model separating the reptilian brain (responsible for instinctual behaviors), the limbic system (handling emotions), and the neocortex (associated with rational thought) is overly simplistic and does not accurately reflect the complexity of brain evolution or the interdependent nature of brain regions in processing emotions and making decisions.[7] Instead of being confined to specific brain areas, emotional and cognitive processes are the result of dynamic networks that involve many parts of the brain working together.[8]

We are social animals, and we are meant to share and exchange emotions, and our body is aware of that. As we adopt a more nuanced understanding of emotions, let us go back to this idea of allostasis, which is that energy regulation and "body budget" that we have at all times, governing our actions in a kinetic chain. Just as we are conscious of our physical energy, we are conscious of emotional energy too! Have you ever met people who are emotional "energy vampires"? It could be tempting to get drawn into engaging with them, but our overall allostasis system finds that draining, so it's naturally part of our wellbeing to learn how to preserve emotional energy through budgeting for human "synchrony," the ability to be in harmony with others.[9]

We've looked at emotions and emotion development in detail here in part because I am making a case for mindfulness, wonder, and creativity applied to our emotions, broadening the landscape of how we approach and communicate what we feel. I've led many trainings and workshops based on emotions and how to forge connection and resilience with transformative transmedia, including spatial computing. Among the biggest keys here are self-awareness, compassion, and curiosity, and mindfulness and wonder play big parts in encouraging and fostering awareness, open-mindedness, and empathy. Understanding the science underneath our emotions informs how we approach, engage in, process, and reflect on experiences.

In the coming chapters, we will explore mindfulness and wonder even more deeply. They make a difference because they encourage and allow us to practice nonattachment. Even as we recognize our emotions, we do not identify with them. We can be self-compassionate, nonjudging, and practice more awareness of ourselves and the changing emotional weather patterns we encounter.

With new strategies for ourselves and our teams, we are able to thrive and also see our own emotional range as a tool for understanding rather than a limit. We transform relationships and open up creativity.

For much of my life, both as a child and as an adult, I have lived in primarily non-English-speaking countries. I was often translating meanings and forming bridges between ideas and applications. Language is fascinating to me. Ambiguity and adaptation taught me to be very plastic in my word associations and to also seek to define terms in ways that defy buzzwords and conventions that can confuse rather than create clarity. When it came to "mindfulness,"* I invented the following description fifteen years ago, and it has stood the test of time. We introduced the Three A's earlier in the book, and now we can approach them at an even deeper level, as one of our leadership frameworks related to wellbeing:

Mindfulness Three A's = Aware, Advancing, and Authentic

These "Three A's" are useful and easy to remember when it comes to mindfulness. We can break them down into deeper descriptions:

Mindfulness =
Aware (awareness of self; of others; of senses, context, and perspectives)
Advancing (active, curious, insightful stretching, outward and inward)
Authentic (accepting of self and others without judgment; willingness to see and be seen; humility and humor)

Practicing awareness increases presence.
Focusing on advancing increases growth mindset.
Embodying authenticity increases relational trust.

It's actually all a practice, and the goal is never perfection. It's simply a way to be in this moment, allowing that possibility of each present moment to unfold, and to befriend yourself, others, and the world around you in the process.

---

*Oh how many times have I received that cartoon, often in jest, of a person meditating with a thought bubble crammed full of images and ideas, with the caption "Mind-Full-Ness"? Yes, it's fun and a fair point: it points to a word that might be confusing to non-native and native English speakers alike!

Practicing mindfulness aligns with our larger thriving objectives, and it can be very goal-driven, yet I find it's best when invited rather than forced.

Jon Kabat Zinn has a similar definition, which I divide in my mind into four parts: "Mindfulness is (1) paying attention (2) on purpose (3) to the present moment (4) without judgment." How simple is that, right? Easier said than done. As the grandfather of the mindfulness movement, Zinn says, *"Wherever you go, there you are."*

Eventually, what each of us comes to realize, or at least what I've come to realize over and over in a spiral pattern that never gets old in its lesson, is that *it's not about me.* It's not about me, it's not about you, and it's not about any forms of separation, perceived or not.

Eventually, in all the beautiful experiences, the self or notion of self dissolves. We become a "we"—a body of energy, something grander than each of us alone. We also understand that, in a world that can dissolve its divisions, there's only "us". . . and that "us" can transcend even the definition of human.

I understand this sounds spiritual, and perhaps it is. When I'm talking about awe and wonder, I'm talking about this kernel. When I'm talking about the metaverse and its seven S's as facets, I'm talking about this. When I talk about the goals of human development and collaboration and cooperative systems of interdependence, it's back to this. Back to relational trust. Back to the we and the us . . . and something larger. I do not define that it has to be spiritual, but it definitely involves that greater purpose and meaning and a presence beyond doing, simply a being. That becomes a form of love.

It all seems to lead there. That's been my experience.

Dr. Dan Siegel, a pioneer in the field of interpersonal neurobiology who trained at UCLA and Harvard Medical School, has introduced a term about identity and belonging: "Mwe" (Me + We) to represent a way that humans can approach connectedness. He addresses identity and belonging in his work, combining personal reflections with scientific discussions of how the mind, brain, and our relationships shape who we are.[10]

We might have grown up living lives of perceived separation, each exploring an isolated "I" in terms of identity and trying to find belonging in a world that reinforced this idea that we are solo, independent creatures, all needing to fend for ourselves in competition and also fit in through cooperation. We were culturally influenced to spend our energy competing with others for worth, and we grew to be motivated by forces of fear and loss. Instead, we can expand our notion of self from an isolated and separate "me" to a relational "Mwe," in concert and synchrony with other humans and the world around us. This wider perspective serves to expand our self-reference, revealing that who we are can be something much more than our brains, our minds, and our bodies—who we are is expansive, intra and inter-connected, and awe inspired. We are creatures who are filled with wonder and built for imagination and connection.

These ideas underpin a lot of why I wrote *Designing Wonder*, a book about spatial computing design that prioritizes wonder and awe. I spent time mapping our relational awe and its ability to move us into transcendent states of collective effervescence. I describe in detail ideas about wonder, which plays a tremendously powerful role in digital wellbeing, in Chapter 3 of this book.

I cannot say it enough: digital wellbeing is a relational, wonder-infused concept, and it is built upon imaginative **layered spatial experience,** by design. It's invitational, more about what experiences you invite into your life design, and should not be approached from a mindset of limits and restriction. Sure, some boundaries and timeboxing are healthy, and the true digital wellbeing promise is one that focuses on the imagination it offers.

On that note, let us look at what I have identified as the Imagination Index: Eleven Keys to Wonder by Design. These are keys for leaders and designers of the future. It's important to emphasize here that every one of us is a creative individual, an artist, and a designer. We need these skills to thrive. It's imperative that we learn how to unlock our imaginations, for the ingenuity of our future professional roles within organizations, and for the flourishing of our whole lives. New KPI's should be built using the Imagination Index as measures. We can all play a part in leading and designing new, abundant futures of thriving, for ourselves and others in relationship. Now is the time to pay attention to these skills and habitudes, encouraging development of all of them. The "how" is embedded in this book, in practices and examples. It's a mindset, skillset, toolset approach.

We could also call the 11 keys this: "Why powerful experiences are awesome; what digital wellbeing should do by design." Wonder and imagination are so richly important that they underpin everything. Here is how to design for them. Remember the first tenet: "Designing for other's imagination depends on you designing for your own imagination."

The Imagination Index: Eleven Keys to Wonder by Design

The Imaginative Leader and Designer . . .

- Sees beyond the separate "self" to a collective "we"—call it a "Mwe" if preferred—that also incorporates integrity of identities. We contain **Multitudes.**
- Understands and embraces the **Fun Factor**. Fun and play motivate everything!
- Builds designs that are **Beyond Simulations** (not just a learning/task practice that replicates physical reality, there's something more at stake and a whole world of imagination to lead us there).
- Prioritizes **Wonder and Awe** and creative reflection and reverie.
- **Celebrates Nonattachment** to typical constraints:
  - Of time
  - Of space
  - Of identity
- Co-creates **Psychologically Safe** environments in which to take risks safely, fail often, and iterate upward.
- Gives **Agency** to self and others—to all!
- Recognizes **Humanity** in self and others—in all!
- Safely embraces challenges and also knows when to relax and rest, in line with a sense of **Flow.**
- Allows participants in the design to *exit mindfully*, and invites the return (**Spirals** of design; nonaddictive algorithms).
- Creates **Social Connection** opportunities; mirror neurons in mindful neural networking fueled by synchrony. This positively impacts belonging and combats loneliness.

It's important for the Imaginative Leader and Designer to use wonder as a pathway to the higher plane of thriving, operating using these 11 keys as a priority. When worlds are built using this code, what will result will feel like magic. It's intentional.

Notice that digital gaming that works well for wellbeing fulfills all of the 11 criteria. It's not a *prescriptive* framework. It's about priorities and being thoughtful and intentional in design. We will discuss how this plays out (ha) when we address games and spatial computing further.

A final framework that is critical to consider when approaching digital wellbeing through a new lens of wonder is the Seven ThEmes of Wonder-rich Spatial Computing Design, or "Seven ThEmes of Spatial Design" for short. I sometimes call them the "Seven ThEmes" or "Seven E's" because they all happen to start with the same letter—how catchy. With exponential advancements in technology and programming quickly ushering us into this next era of spatial computing and design, empowered and animated by AI, it's imperative that we pay attention to these factors and "get it right" ethically. We need to first understand what each facet means and use our best scientific knowledge to inform our designs and actions.

Seven ThEmes of Wonder-rich Spatial Computing Design

- **Experience** (doing it)
- **Emotion and Empathy** ("in feelings" we encode)
- **Engagement** (how to invite play; distraction from pain)
- **Entrainment** ("internal rhythm" or flow of experience)
- **Exercise** ("exergaming"; exertion)
- **Embodiment** ("really there"—full body + transformative)
- **Expression** (story; memory palaces)

The Seven ThEmes are key layers that all contribute to meaningful, moving spatial computing design, useful for game considerations, and all sorts of learning and media applications. These can be the core principles that designers use when approaching spatial activations. We will explore these more and describe them in detail later on in this book.

Digital Wellbeing Through a Lens of Wonder Frameworks in Summary: Four Culture Cornerstones; Three A's of Mindfulness; the Imagination Index; Seven ThEmes of Wonder-rich Spatial Computing Design

As we move to application examples in future chapters focusing on gaming, extended reality, and spatial computing, we will keep these digital wellbeing frameworks top of mind: the **Four Culture Cornerstones** (Dignity, Freedom, Invention, and Agency); the **Three A's of Mindfulness**; the **Imagination Index:** the **Eleven Keys to Wonder by Design**, which are habitudes or ways of being that a designer must prioritize to elicit wonder in their creative environment and its products and experiences; and the **Seven ThEmes of Wonder-rich Spatial Computing Design,** also known as the "Seven ThEmes," which will pave the way for the future and must be prioritized in all of our designs, along the full continuum of applied technology with wellbeing as foundation and priority. These frameworks are our essential touchstones, reminding us of the intention that drives every practice.

Personal Reflections

1. When you initially thought of the terms wellbeing and digital wellbeing, before reading this book, were there any connections to boundary setting that came to mind?
2. How does an intention to use approaches to technology to open up creativity and wonder change the narrative surrounding wellbeing?
3. What do you think about the new research about emotions? In what ways is it liberating or thought-provoking?
4. When you look at the Three A's as approaches to mindfulness, which one has made the biggest impact on your life, and how?
5. What is your previous experience with spatial computing? What are you curious to learn about and explore more related to the topic?

Reflections for Leaders

1. How could adopting the Four Culture Cornerstones transform the quality of creativity and wellbeing for you and your teams?
2. What have you encountered in this chapter that challenged a previous way of thinking or reached you in some new way?
3. What aspects of the Imagination Index are you and your teams already practicing? Which ones would you like to implement, and how could you envision going about that?
4. What areas of the Seven ThEmes of spatial computing are you looking forward to exploring?
5. How does adopting more of a "Mwe" mindset change the nature of the workplace and business?

Extensions for Future Exploration

I have always been fascinated by quality of life indexes and measures for success. When I lived in Zürich, I cofounded an organization called the Center of Wise Leadership. We were looking at organizational models that were adopting new measures, given the rise in automation and AI to help increase efficiency. How do we measure quality of life, and how do leaders now adjust to a world that is increasingly "VUCA"—volatile, uncertain, complex, and ambiguous? Change is not going away, and it's how we adapt to that change and reach out to support each other as an interconnected system that determines our overall resilience, individually and collectively. As an extension to this chapter, I would choose to look at even more use cases and frameworks for wonder, imagination, and digital wellbeing across different sectors and disciplines worldwide. I would spend time interviewing and researching with communities, and take a holistic longitudinal approach. I strongly believe that digital wellbeing needs to have a dynamic, context-driven outlook, with new KPI's for business that reflect those values and measures, and it's important to deeply investigate systemic influences as we frame new future infrastructures for thriving as individuals and in organizations. In the following chapters, you will see my work addressing different cohorts and generations. We will investigate topics associated with digital wellbeing, such as gaming, spatial computing, virtual reality, wearables, movement and exercise, and relationships. This book contains many dimensions, and it's a tool for conversations and applications that can extend and transform quality of life at local and global scale.

3 Facets of Wonder: Implications for Digital Wellbeing

This chapter is the "bead on the necklace" of our digital wellbeing experience that addresses wonder, awe, and imagination first and foremost. One could argue that human imagination is the greatest asset for humanity. Our resilience, empathy, problem-solving, meaning making, relationship bonds, and sheer survival depend upon our ability to thrive with imagination. This next section will address the "how" of building and flexing our creative muscles and also investigate many forms of digital media that can enhance our imagination, not threaten it.

Along the way, throughout this section especially, I encourage you to keep a journal nearby, digital or analog, where you can sketch notations and record personal reflections and experiences that might come to mind. You could also visit my website, caitlinkrause.com, where there are companion resources and interactive exercises that accompany these ideas.

What is awe? What is wonder? How do they relate to imagination and wellbeing?

In the beginning of this book, we imagined an experience with awe, a powerful emotion that has the opportunity to completely transform our experiences and mindset. Now, we are returning to the topic to understand more about awe and its complementary emotion of wonder, learning more about how they have such transformative properties and can impact wellbeing.

I'm especially interested in "collective effervescence," which involves shared experiences in awe. This term was originally coined by the French sociologist Émile Durkheim to describe the sense of energy and excitement that arises when people come together in shared experiences.[1] This takes place in many shapes and forms, from dance celebrations to sporting events, cinematic cheers to religious ceremonies. Humans need rituals. We feel more awe when experiencing something moving together.

Dacher Keltner, author of *Awe*, recognizes the astonishing nature of the emotion. Keltner, a professor of psychology at the University of California, Berkeley, says awe is "the feeling of being in the presence of something vast that transcends your current understanding of the world."[2] He cites collective effervescence as one of the top ways humans experience awe.

Later in this book, when we talk about awe-inspired design for spatial computing and immersive virtual reality experiences with flow states, we will animate collective effervescence even more deeply.

Wonder and awe have many benefits for wellbeing. The two concepts, while closely related, have distinct values in both personal and professional contexts. As we have experienced and talked about, awe is an emotional response to perceiving something vast that expands our way of thinking and feeling about the world, often leading to a diminished focus on the self and an increased sense of connection with others and the world at large.[3]

Awe is an emotional response to perceiving something vast that transcends our current understanding of the world, often leading to a diminished focus on the self and an increased sense of connection with others and the world at large.

This shift in perspective can have profound implications for wellbeing, encouraging prosocial behavior, fostering a sense of belonging, and promoting generosity.[4] In professional settings, experiencing awe can inspire a collective vision and purpose, enhancing team cohesion and motivating individuals toward shared goals. The sense of being part of something greater than oneself can also lead to a more fulfilling and purpose-driven work life, contributing to overall job satisfaction and emotional wellbeing.[5]

Wonder, while similar to awe, is characterized by a heightened state of awareness and curiosity sparked by something surprising or delightful.[6] It drives the desire to explore and understand, fostering an environment of creativity and innovation. We're going to investigate concepts surrounding wonder and awe more deeply in this chapter, including some of my previous research on the topic that contributed to my 2020 book, *Designing Wonder*.

In personal contexts, wonder can break the monotony of daily life, encouraging exploration and the discovery of new interests or passions. Professionally, it can stimulate creative problem-solving and innovation, as individuals are motivated to look beyond conventional solutions and explore new possibilities. Wonder also plays a crucial role in learning and development, as it encourages open-mindedness and adaptability—qualities that are increasingly valuable in the rapidly changing landscape of the modern workplace.

Wonder, while similar to awe, is characterized by a heightened state of awareness and curiosity sparked by something surprising or delightful. It drives the desire to explore and understand, fostering an environment of creativity and innovation.

Both awe and wonder contribute significantly to wellbeing by enhancing emotional resilience, promoting mental health, and cultivating a sense of connectedness and purpose. While awe may lead to a more collective and purpose-driven perspective, wonder fuels individual curiosity and creativity. Together, they offer a comprehensive approach to improving wellbeing in both personal and professional spheres, encouraging individuals to engage more fully with the world around them and to pursue more fulfilling and meaningful lives.

As we look at facets of wonder even more deeply, we can see similarities to awe and a foundation in mindfulness. Emotional states of wonder and awe can be consciously practiced. Significantly, such states can contribute to enhancing the abilities to design incredible immersive applications and foster a better sense of wellbeing as we give context to these traits.

Facets of Wonder[7]

> *Where the imagination is alive, wonder is completely alive.*
>
> —*John O'Donohue*

Wonder is a feeling, and emotion, and also a sense that we can tune our appreciation for. In ways, **it's a capacity that is a primer and prerequisite for a lot of other things to come about**. Imagination, mindfulness, and wonder are intertwined in powerful ways. Here, because experiences involving and inducing wonder are so meaningful in spatial computing and particularly VR, we will look at several different facets of wonder. In each segment, divided by facet, I offer new ideas and approaches to concepts, experiences, and applications in VR and in day-to-day life.

The theme of wonder traces its roots to times long before Lewis Carroll and Wonderland captivated our fancies and fantasies. What can it mean, and how does the concept have facets that transform what we experience as "reality"?

Facet One: Wonder as Intentional Attention

Intentional attention has a lot to do with the mindfulness practices involving awareness that we explored earlier in the book. How does deeply attending to the "now" moment lead to wonder?

On any given day, thousands of creative ideas flood our brains, often in a moment-by-moment succession where if we do not make space for wonder, we might dismiss them because we have seemingly more pressing thoughts to attend to. We might wonder about the way a bird landing on a lake seems to skim the water for several hundred yards, preferring to fly close. Is the air draft better there? Is she hunting for food beneath the surface? Later, we might investigate why. For now, we can just watch and appreciate the noticing.

This intentional attention practice has certain conditions that serve to make it special: it's not fixed; it's not strictly purposeful. Because we do not *expect* anything to come out of it, what ends up happening is a certain magic of "being there" in the moment.

This level of wonder is tuning into, and attuning to, the senses. It's a rich sensory experience, to direct our attention in this way, and to honor the art of noticing. I would say this also might be why humans love traveling so much. When we travel, we are surprised by the newness of what floods our senses: the sights, smells, tastes, sounds, textures of this new environment, so different for us to experience, so invigorating. We notice those details and *pay attention* to them because they are so different from the patterns we have grown to expect. What stands in our way from seeing with new eyes, and using intentional attention, on a daily basis? The obstruction usually involves our own mindset. The thread of wonder needs to be prioritized.

Time and safety are precious resources. For all of these wonder-related facets to be involved in an experience, they need each of us to purposefully set aside the time and prioritize the practice and the experience. We must feel safe in order to do this. These are not superfluous requisites. Wonder is the base of letting go and being able to transcend limits.

Facet One, *Wonder as Intentional Attention*, depends upon adopting qualities of mindfulness and forming them into a mindset. We can apply the three mindfulness A's of awareness (of the present sensory moment), advancement (celebrating growth and stretching past perceived limits), and authenticity (truth without judgment) to help prime ourselves for wonder. Qualities of mindfulness and wonder are linked.

When we practice mindfulness, we begin to place our attention, with great intention, on what naturally emerges in the moment. We can respond with curiosity, openly wondering in a way that can involve both asking questions and simple, pure appreciation.

This facet can involve asking open questions, among them, "What if . . . ?" and "How . . . ?" It can involve looking at something very deeply first, and then making a creative connection or parallel that leads to a metaphorical wondering, for example, watching bees circling around flowers, drinking up nectar, pollinating, hovering in a certain way. We do not have to know all about what they are doing, or even think about how to explain it, to get a sense of wonder about it. We drink in the act of deeply noticing it. We absorb sensations associated with being in full presence to appreciate the wonder.

When we approach this in a digital landscape, this facet of wonder involves releasing the analytical brain in favor of absorbing the purity of experience itself. For example, we enter an immersive digital landscape and take in the feeling and sensation of "being there" first, rather than aiming to dissect each element. If I lead an exercise and the first thing someone asks me is, "How did you import these objects?" they are using a different part of their mind, trying to dissect before noticing the quality of the experience. Later stages can involve mechanics and deconstruction to reconstruct. For now, it is "being in wonder" that is the priority.

In sum, this facet of wonder involves placing a priority on process before an act. The act would be the subsequent "action" of what is already held in priority in process. The process involves mindfulness and fixed attention, which is the intentional awareness placed on the natural world. The natural world includes our internal state of being, too. Taking time, and pausing, is part of this natural state. It's appreciation rather than dissection and analysis. **It reflects a necessary paradox: while it takes skill, this level of attention and wonder can be seen as naive.** This could also be why many of us, as adults, resist entering the same states of wonder that came naturally to us as children. Looking open, rapt with attention placed on the sheer beauty of something, this is daring. Wonder reflects a wildly courageous, beautiful state of raw vulnerability. Poetry subverts this and can surprise us, too, inviting us to fall into wonder through the language, the appreciation for an exquisite moment described in evocative detail.

This facet of wonder involves placing a priority on process before an act.

Implications for Digital Immersive Media In a similar way to how poetry can invite wonder, VR can teach us to notice and appreciate the details of a scene, drinking them in with our senses. We have dedicated our time and attention to that space, and it can cause us to have an even deeper connection with what surrounds us. Is this deeper *presence*? I would conjecture yes, because we are away from distractions, and can immediately drop in, intentionally, on the present moment.

For example, I led a training session for a group inside of an immersive social environment in VR. In this experience, I chose a 3D scene, a seascape with cliffs on one side, the ocean stretching out on the other. As we stood on the beach, we took a few minutes to acclimate and practice breathing exercises at the beginning, and then everyone had time to spend in quiet contemplation. They could face in any direction, and the invitation was to notice something specific—the sound of waves, the feeling of your own grounding in your feet, your breathing coming in and out. We had only sight and sound as sensory inputs, yet the other senses were filled in by our imagination. Looking toward the light of the setting sun over the waves, for example, my face felt warmer. I felt lighter and even more buoyant in my process of focused awareness.

Facet Two: Wonder Beyond Ourselves

The process of noticing that we explored in Facet One leads to Facet Two, "wonder beyond ourselves." When we are tuned into our senses and the power of observation, appreciation, and presence, we recognize our own level of perception as coming from a limited vantage. We also recognize the vast perspectives that exist beyond our view. In ways, this has to do with engaging in self-expanding practices that relate to recognizing multiple layers of identity so that we can recognize our own diversity and richness. As Walt Whitman said, "We contain multitudes," and part of wondering beyond ourselves begins with wondering at our own complexity and forming a sense of how much there is to appreciate, inside and beyond ourselves.

Many organizations have explicitly stated that fostering inclusion and a culture that uplifts diversity are among their core missions. What capacities, intentions, and capabilities are necessary to make this come about? Focusing on an ability to wonder beyond ourselves is a key leadership practice and skill that has the ability to radically change organizations from the inside out.

This is an expansive practice and philosophy that connects to freedom and complements building empathy and compassion for ourselves and for others. We can involve storytelling, authoring our own narratives and using them as a vehicle to express, to connect, and to inspire a sense of curiosity that links to our wonder. As we use this capacity to drop into the moment, we remind ourselves to keep openness as a constant. As W.S. Merwin says, "Don't close off your life."

Implications for Digital Immersive Media VR has been called an "empathy engine" by Chris Milk, and it has been shown to have an immense power in transporting users into another place and time, giving them the ability to see from someone else's eyes, possibly imagine what they feel, and form better understanding. That level of understanding is dependent upon having context and applying meaning through the experience.

For VR to truly encourage and empower a user to not only feel a certain way and experience empathy but to take action to make the situation better—the working definition of "compassion," differentiating it from empathy—users must inhabit that place and space, focusing their attention on the VR environment they are experiencing *and* allowing for a breadth and depth of reflection. They cannot simply be engaging in an experience of passive entertainment; they must interact with the materials in a context-driven way in order to have an appreciation for what they are perceiving. At times, they must even lose their own self-attachment in order to perceive situations from another viewpoint. There is a backstory; there is something at stake. This takes time and care to craft and also to execute.

Jeremy Bailenson, founding director of Stanford's Virtual Human Interaction Lab (VHIL) and author of *Experience on Demand*, said, "We often get asked 'Does VR cause empathy?' And I would never talk to a reporter and ask, 'Does the medium of video cause empathy, or does the audiostream cause empathy?' The answer is clearly that it depends on what you do with it. For a long time, we thought about the medium but not the content. This sounds really obvious . . . but think how long it took for film to evolve. . . . In general, when you compare VR as a medium to a videostream, or even to the gold standard for empathy 10 years ago, which was mental perspective taking . . . in general virtual reality caused more behavior change, and we think that VR is the mechanism behind that behavior change, compared to the other media."

One of the areas VHIL looks at involves body language, physical movements, and learning outcome predictions. Content in VR matters, and a user's engagement with that content is paramount. They've got to do the work and become actively involved! It's not passive. And what is a great measure of "the work," aka engagement, especially when it comes to empathy? *Movement.* The physical movement could be an indicator of internal emotional movement, too. If a user does not show movement effort, the stats in studies do not show any difference in levels of care or empathy. If a user is attentive and engaged in the live moments of the experience (indicated by *body movement!*), that's where we see a significant shift.

The wonder of multiple perspective taking is related to body positioning and movement. The appreciation for other perspectives, including elements of empathy, is what Facet Two is all about. It is certainly possible to learn how to practice perspective shifts and empathy, and it's powerful when it happens. It can change leadership by increasing levels of presence, humility, and curiosity. We prioritize aiming to understand others first, knowing that their experiences will be different than our own. We can all seek to reach others where they are, and to form a meaningful connection. I think of the adage: "How can I reach you if I don't understand you?"

Danish philosopher Søren Kierkegaard (1813–1855) famously said, "In order truly to help someone else, I must understand more than he—but certainly first and foremost understand what he understands. If I do not do that, then my greater understanding does not help him at all . . . all true helping begins with a humbling."[8]

This is a quote that is all about the "Wonder Beyond Ourselves." Increasing understanding of others and truly developing the ability to see beyond ourselves is a top leadership skill, useful for organizations and education alike. Because VR is a place where different stories, experiences, and perspectives can be more accessible and customizable in their design for specific needs and use cases, it's a powerful environment that can enhance significant transformation in mindset, relationships, and culture. The separate self dissolves, and what remains available is pure connection and understanding.

Later on in the book, when we talk about spatial experiences, we will look in greater detail at how immersive experiences, including VR and gaming, can support wellbeing for individuals and teams collaborating, to increase social connection, emotional intelligence, imagination, compassion, innovative problem-solving, and more.

Here, we lay the groundwork. There are also exercises online at caitlinkrause.com for those looking to practice Facet Two.

Facet Three: Surfing Wonder

When I was very young, up until age four, we lived by the ocean. One of my favorite things to do was walk on the beach. It became a "wander and wonder" exercise, exploring tide pools, investigating miraculous horseshoe crabs, and jumping waves as they rolled in and pulled back out to the sea. As I grew older, I would body surf in summertime, and I enjoyed the feeling of being in motion with the waves. I use a metaphor of surfing in my meditative practices, and I often ride the waves of my thoughts in a way I associate with wonder.

We call this facet "surfing wonder" because it has everything to do with flow and connection. It's about being present in the environment and moving with it, in consonance, much like a surfer in the ocean learns to feel the wave under the board and move with it, using its momentum and making micro-adjustments when needed. A mind-body connection and a certain proprioception occur, where the body is able to perceive itself in time and space and form a deep level of knowing that goes beyond typical thoughts and cognition. It's a new level of referencing.

There are two important points for consideration: First, this proprioception of self-perception in time and space does involve cognitive processing, and it's a healthy practice for the brain to engage with. The part of the brain that controls proprioception (consciously and subconsciously) is the cerebellum, the same part that is used to affect balance. I use it all the time in sports, in my yoga practices, and in my intentional movements to reinforce a mind-body connection.

Second, we can think about two traditional major types of proprioception as we approach this notion of surfing wonder: egocentric and allocentric reference frames.

An **egocentric frame of reference** basically means that objects are recognized relative to the perceiver. It is more of an individual view.

An **allocentric frame of reference** refers to objects that are represented relative to the environment that is outside of the perceiver—their surroundings, others. It is a more collectivist view and can involve the ability to recognize and appreciate object-to-object relationships.

Some criticize modern technology culture, saying that there is an egocentric approach to a lot of applications. When we look at nature's examples of meaningful cosensing and collaboration, and have more of a biocentric approach, we see many examples where allocentric referencing produces a more resilient, thriving system of belonging. This naturally solves the loneliness crisis too. Part of the intimidation of such new frames and systems of reference is the fear of loss of control. What will happen if we learn to adopt new frames of thinking where we are in service to a bigger picture, where we are not center of the universe?

There are many ways to explore allocentric spatial processing topics more deeply, and research is actively taking place right now (take Harvard's imagery lab, for example), looking at how mental models can be represented and explained in different interactions and case studies.

The brain is constantly forming mental models of its surroundings and its world. We build models to understand a context in our surroundings. We form schemas and patterns so we know what to expect, and when we receive surprises or new information, our brain encodes new understandings to accommodate the new information. It's a fascinating process, and what it also means is that honing flexibility of mindset and models does not come naturally. It takes training. Meditation can become an integral part of developing self-awareness and also the ability to recognize our own mind and its reference framing. I would conjecture, from my own experience, that it also allows us to move fluidly between egocentric and allocentric referencing, and even transcend them when we are "surfing wonder" and engaging with other facets of wonder in a certain way.

To understand this, let me use a few examples of what surfing wonder can look and feel like. It usually does not just happen by chance. You build up to the experience of surfing the wonder of your thoughts by prioritizing certain preconditions. First and foremost, you need to feel safe, because safety is a key prerequisite to the brain's ability to "let go" and ride the ride of our train of thought. If we are not feeling safe, our limbic system in the brain triggers a response mechanism that overrides the creative processing. We're not able to reach higher levels of brain functioning, such as creative ideation and "flow state," because we are in survival mode. This is a necessary thing and actually good that our brain has this trait of recognizing and responding to perceived danger, because it's what successfully ensured our survival throughout the ages, even dating back to the times when we would need to respond quickly to actively run from threats. When we feel unsafe, our stress triggers an immediate response, in which cortisol floods our system and we are able to feel that rush that will lead us to fight, flee, or even freeze in some cases, if it serves our purpose.

The trouble is, these days, we are still hardwired to experience that flood of stress response when it is unnecessary and even harmful. It can become rather addictive, to adopt a toxic relationship with news, work, and other perceived expectations that might cause us to feel we are clenching, reactive, and lacking control and agency. We might also have unwittingly embraced a belief that to show that we are dedicated, passionate, and hardworking, we need to exude a demeanor of frantic, always-busy, always-responsive chaos and pessimism. It's a culture of overwork and cynicism that had been a workplace trend until very recently, when many companies witnessed their employees experiencing burnout.

Now, new solutions for overall wellbeing invite us to form a new relationship with the workplace. These methods are just starting to be explored and prioritized—hence, this book! Mindfulness has been part of this movement, and I see a cultivation of spaces and places for wonder as integral to this process. How can we increase safety and trust so that we can comfortably "let go" and surf wonder when we feel the invitation?

Surfing wonder, then, is about letting go and following the wave of ideas, without even labeling them as thoughts. You ride the wave, so you feel it instead of identifying it. We already know that our cognitive brain functioning wants to create patterns and then build meaning from constructs. Here, you use a "Surfing Wonder" process to recognize that tendency and construct and simultaneously break it by allowing it to dissolve. You are so in tune with riding that wave that you can quiet the mind that wants to analyze at the same time. For a moment, perhaps at the beginning of this exercise in surfing wonder, you are conscious of your own limiting narration of the experience taking place, and you then let it float away. It's a gentle practice, much like mindfulness meditation, where the censors are off, so that the thoughts that come are what is leading. They are the wave, and we ride them, trusting and feeling our balance.

Poems as Resources and Entry Points for Surfing Wonder:

"The Inevitable Lightness" by W.S. Merwin

"The Wings of Daylight" by W.S. Merwin

Implications for Digital Immersive Media Extended reality, and especially virtual reality, where the external world is completely replaced, involves worlds that already invite this type of surfing. When people become used to indulging in this state and entering into an appreciation for what it's like to surf the waves of the mind's experiences, each time in VR can feel like an open discovery that leads in a direction of deepening understanding that we could not necessarily pin down or predict in advance.

Remember when, earlier, I talked about the prerequisite to first feel safe in order to then experience surfing wonder? Well, this is *exponentially the case* when it comes to VR. Not only do we have to feel safe entering the VR experience, but we must feel continuously safe and actively cared about while we are present in a virtual space and experience. What does this mean? The mentor carefully sets up the conditions and limits that ensure our comfort, our dignity, and our very essential belonging.

The "SCUBA model of VR Experience Design" is a methodology I invented that encourages and empowers a mentor to bring wonder about in different ways. It can certainly involve elements of surfing wonder. Nearly any content can fit into this space. The essential part is about those values and behaviors that will allow anyone entering that virtual space to have that basis from which surfing wonder, as well as the other facets, can naturally take place.

The second prerequisite, in addition to safety, is *time*. For this VR experience to breathe on its own, and for surfers to be able to "ride the ride" and engage deeply, they will have to dedicate the time to the experience. This sounds simple, but it is hard to achieve in the modern world. Ideally, they avoid having another place they need to be immediately before and after the experience in VR. When they have defended and carved out time, they feel unburdened by any sense of urgency and pressure from the outside world. What a luxury, right? It's not a nice-to-have. It's essential.

John Carmack said onstage in 2019 that he was looking forward to the ability to receive text messages and alerts inside of virtual reality. I cringed when I heard him say that. Did he really mean it? I still wonder. I protest: the ability to be always available quickly translates into the ability to be always *interrupted*. We need safe, dedicated spaces and places to experience uninterrupted wonder. VR can be that beautiful space. By now, to many of our dismay, the technology has already evolved to allow you to be "pinged" inside of VR at a moment's notice, in which case I would urge you to resist the dopamine-related addictions to those rings and pings and turn notifications off. It's the only way to surf wonder (inside and outside of VR) in a way that is genuine and beautiful. You need to ride it, with the ultimate freedom that is part of the experience, and your body will know that flow when it gets there. Dedicated safety, time, and space with the inner landscape are part of this.

I know there are exceptions, but I feel strongly that there is a place and time for uninterrupted wonder. As a writer, I know the practice of patience, and of perseverance too, and it seems as if the best inspirations come when my mind is caught "wandering and wondering" and I do not have a sentinel on alert notifying me about where I have to be. I lose myself. If you are reading this and you think, but what about my family, my job, my leadership position, my captaining of the team? How will the world go on without me? Well, I hear you, and these are real concerns. Cal Newport and other productivity researchers, however, counter that the time spent in *Deep Work*[9] is where imagination and innovation happen. More on that to come in Chapter 6.

Surfing wonder reminds us to connect with the moment-by-moment unfolding of the present moment, truly living in the now. As Marie Howe says, this is "what the living do," and this facet needs to meet you where you are at. So, one strategy, ironically, is to set a timer for yourself and block off any dedicated time you have—10 minutes, 30 minutes, 1 hour—to strategically, mindfully let go and surf wonder (this can also work for meditation or VR practices in general). Do not be afraid or ashamed of these practices. Deep work and wonder-rich creative flow support our minds being able to focus.

Along with that strategy, I would add: Try to make your practice in surfing wonder visible to your close friends, loved ones, and work collaborators. It's not something to hide, though it might feel very personal. The reason I say this is twofold: it allows you to fully embrace the expansive qualities of the practice, and it also ensures that your intention is clear and visible up front. That way, it's understandable when you do not, for example, answer your text messages and emails all the time, especially during certain hours.

Do you remember having friends who resisted getting a cellphone when the mass adoption first happened in the late 1990s and early 2000s? I envied my friends who refused to make "the switch" and move into the era where we were "always reachable." It affected our social behavior and our friendship in a significant way, as there were different expectations for friends without cellphones. We both had to show up in a place and time when we pledged to, with a certain reliability. We had to make time for each other. It encouraged us to connect in-person and to allow for the longer periods of disconnection, when we might be out of contact but not out of each other's thoughts. A sentimental statement, but it is true in a very practical sense. Maybe the disconnection was healthy for the relationship. There is irony in a book about connection using technology noting that one of the natural components is the benefit of disconnection.

It's natural to be unreachable, at certain times, for certain purposes. We do not need approval from family and friends for our surfing wonder plans. We entrust them with knowledge so that they can manage their expectations. In addition, we can always use a setting to allow certain important calls through when we are in "do not disturb" mode. This can ease our mind and make the surfing easier.

A Note on Brainwaves and Brain States As we contemplate the first three facets of wonder: *Wonder as Intentional Attention, Wonder Beyond Ourselves,* and *Surfing Wonder*, we might notice that all the facets have a certain flexibility of mindset, able to lose oneself in midst of the experience, or follow a train of thought without forcefully directing it. Well, what's happening here, essentially, is related to brain states and brainwave functions. Scientists have been studying the states of the brain for years. We might need to "think" in a logical way about inviting ourselves to enter a state of wonder, and then the part that involves riding the ride relates to brain states and processes that are less associated with conscious thought.

It very well could be that surfing wonder, in practice, has to do with a combination of alpha and theta brainwave states. Alpha waves have been associated with creativity in recent studies too.[10] So, there's a mix of interplay here. In fact, some combination of all of the states, at different times and frequencies, are present during experiences involving wonder. Understanding more about our brainwaves can serve to inspire us to "feed our brain" with a variety of conditions and inputs.

Facet Four: The Overview Effect and Awe States

The notion of the Overview Effect is not a new one; astronauts have been experiencing elements of it for decades, as they look down on Earth from space and experience a sense of awe, both a largeness and a smallness all at once, as well as a feeling of great interconnection as a planetary society.[11]

We've likely experienced something like the Overview Effect when we are lying on our backs at night, staring up at the stars, having deeper thoughts about the galaxy and our place in it.

This feeling itself is the Overview Effect, a cognitive shift in awareness and consciousness. Philosopher and aerospace researcher Frank White coined the term in 1987, detailed in an article by NASA.[12]

Much of the awareness that contemplating Earth from this "overview" level inspires is an appreciation for the fragility of the "pale blue dot." We also form a recognition that many of the boundaries and divisions created by humankind—divisions between nations, societies, etc.—dissolve when the planet is viewed in its wholeness.

Take this quote by Apollo 11 astronaut Michael Collins:

> *The thing that really surprised me was that it [Earth] projected an air of fragility. And why, I don't know. I don't know to this day. I had a feeling it's tiny, it's shiny, it's beautiful, it's home, and it's fragile.*[13]

When we contemplate the "Overview Effect" facet of wonder, we step back (or rise up) in consciousness to embrace a new perspective that reflects awe and recognition. We get very quiet; we become reverent. We feel our humanity and our fragility. We also feel the power of possibility and a sense of belonging.

Some of us might have experienced the Overview Effect while on a flight, looking down at the towns and cities, maybe seeing the connectedness of the patterns, and the beauty of mountains, rivers, forests, and lakes from a different perspective.

Many modern leadership programs encourage participants grouped in small cohorts to question assumptions as they open up to more humility, more genuine appreciation for interconnection. They start to notice this "oneness" view playing out in approaches to organizational leadership and strategy development, as there is a sense of belonging, even in VUCA (volatile, uncertain, chaotic, and ambiguous) worlds. In the midst of complexity, with all of the beautiful diversity and differences, we are not separate.

How can we use this awareness, this wonder and awe, to engage with our colleagues differently and to approach our work and sense of purpose from a new scale? Remembering oneness also reminds us that we are each less alone than we might think, and we can do more together, by design. Communication improves, and there's a reverence for the larger goals.

This facet of wonder also carries with it a connection to our home planet and greater dedication to working to protect the environment. We see that we were made for the world, and it was not created to be consumed, but to be protected. This is the type of genuine humility and connection to nature that the Overview Effect inspires. Certain poetry also carries this message of the Overview Effect of oneness. Some of these awe experiences have made their way into my poetry book *Digital Satori*, and there are many ways to encounter transcendence and a feeling of deep connection through art and science. More examples and references are online at caitlinkrause.com.

Implications for Digital Immersive Media Implications for immersive media are as immense as the universe itself, and that's an understatement. I've given a talk on the moon in ENGAGE's immersive social VR platform;[14] I spoke about the Overview Effect as we were experiencing the Overview Effect ourselves. Being in immersive spatial environments invites a greater state of awe and sense of flow quite quickly, and it has something to do with the implicit and instantaneous consciousness we adopt that we are inhabiting shells, whether our physical one or our virtual extension of avatar, that become our vessels. We are contained and uncontained, when it comes to the expansiveness we are part of, and floating in a virtual galaxy helps usher us into a new state of wonder.

VR and other immersive media offer a chance to be quite literal about inducing the Overview Effect. We can also access this feeling anywhere, anytime, using our imagination as the conduit. Contemplative exercises can bring up this wonder and awe feeling in different ways, as we continue to explore. How will you be inspired to include the Overview Effect and greater sense of wonder into your experience design processes, for you and others?

Facet Five: A Hidden Wholeness

Now, after having approached wonder from different angles, seeing facets come alive, we might feel as if things themselves take on new meaning in relation to each other. We gain an overall sense of something instead of trying to break down the parts. Poetry encourages this, through tuning into a moment. A part stands for the whole. Synecdoche. Greater than Sum of Parts. Enso and Beautiful in Broken. Yin and Yang. Suffering. Joy. Completion.

All of these elements play a part in the wonder of the whole, which is about recognizing the (sometimes hidden) wholeness of things.

This facet takes part of its title from author and philosopher Parker Palmer, who wrote a book called *A Hidden Wholeness* in which he offers ways to "combine soul and role," offering ways for us to "live undivided lives—lives that are congruent with our inner truth—in a world filled with the forces of fragmentation."[15] Recognizing our own wholeness is part of gaining a greater sense of what is genuine and authentic and whole about the interconnectedness of the world.

How do we make the invisible visible? When Tom Furness, the "grandfather of virtual reality," and I were talking about this facet of wonder, I described the sense of wholeness and the knowing that it all fits together in some cosmic synchrony of recognition. He brought up the gestalt. When he mentioned it, I had a vision of seeing one element and realizing the other naturally fitting elements in a puzzle, without a need for deconstructive analysis. Tom said, "Everything in a moment contains the whole of the past, and it becomes quite clear." The symbolic wholeness and clarity serves its completion, and we have a sense of recognition, belonging, and knowing. Poets and artists tend to be skilled at capturing a gestalt. A small, distilled element is able to convey deep meaning and reflect the intention of the whole.

Gestalt itself is a German word, and like many great German one-word complex philosophical concepts, there is no exact translation in English. It loosely means that the whole is greater than the sum of its parts. You cannot gain an understanding of a system as a whole by examining the loose collection of individual parts, although I do find that understanding the context and differentiation of the parts encourages me to honor even greater synchrony and diversity of collaboration when they operate as a whole.

The notion of wholeness has a great deal to do with human perception. Humans tend to think in patterns. We sense relationships, and we apply and ascribe a sense of meaning to what we consider a "configuration" or pattern. We pay attention to relationships. Take listening to a melody, for example. We listen to it as a whole and appreciate the composition as something "more than the sum" of the sensory components. There's an added quality to what happens when the parts are played in sequence, all together, and the philosophers would call this (naturally!) "*Gestaltqualität*" or "form-quality." I think it has to do with an organism's ability to perceive aesthetics in some senses.

We enjoy collages, similarly, for the freedom of the individual parts and the assembly into a whole, often with whimsical elements that surprise and delight. Examples of this wonder of wholeness are absorbing the perceived randomness, changes, and complexity that also reflect belonging and, somehow, organization and coordination. This awareness brings us wonder, and it rises in our everyday life in a moment, often ephemeral. It can happen when we encounter a scene and form a pattern that naturally tells our body how to behave and act within the system.

Priming ourselves for this form of wonder and awe can be transformational. We seek continuity and cohesion, and here are natural cues informing us, sometimes without our conscious response. This can be considered a perceptive phenomenon when it occurs and might be experienced in the way we would describe having an insight or intuition about something.

Implications for Digital Immersive Media Everything contained within an immersive experience is chosen and crafted with intention, and the combination of specific elements is what causes an *organization*, a wholeness that the participant or user can feel and respond to in a way that is rich, resplendent, and transporting.

When I'm in a well-conceived and well-executed experience, I feel as if I have entered another dimension. I am connected, complete, whole. This reflects the hidden wholeness of such wonder.

One of the memorable times that I felt this was when I was part of a concert captured volumetrically in the Wave. When the music began, I immediately felt connected and part of the show in a way I had never experienced before. The music had a large part to do with it, but it was the immersive landscape, the darkness, being out in a field, then the light as well as sound, the feeling of completion, and the dissolving of anything else, just the melody and the wholeness. You can watch a 2D video capturing this immersive 3D experience and feel a shade of what it's like.

When I was part of the live concert, I lay down and stretched out my arms, making snow angel arcs on the carpet (grass in VR!), moving in a way that felt like freedom and meditation. I lost any inhibition, I felt safe, and I knew that the whole scene, the images, the sounds, the vibrant sparks, the scenes and nature, they were all a giant swirl of belonging, and I was there too. I did not disappear. I was fully surrendering to the moment, and I also became an integral part of the experience by feeling a deep connection. It was emotional, pure, and related to awe, wonder, and alignment. I believe that the closest word for how I could reflect this is *consonance*.

Part of the reason that these five **facets of wonder** that we have explored can be so intensely remarkable is that they have the capacity to completely transform the quality of our experiences in ways that are surprising, astonishing, and body-and-mind shifting. Knowing about each of them, accessing them and incorporating them into our priorities in our personal and professional lives is of great benefit to everyone. The sections labeled "Implications for Digital Immersive Media" are there to make the connection with technology explicit. So often, wonder and awe are seen as a connection to physical nature. While this is undoubtedly the case and a wonderful thing, it is not exclusive. We can bring this same reverence for wonder to our relationship with digital technology. Using wonder as a conduit and also using technology as a tool for enhancing mindfulness, wonder, creativity, and wellbeing is the proposition and possibility.

This chapter is meant to serve as an open conversation . . . let us reflect, respond, and keep the dialogue expanding. By combining wonder with our address of mindfulness, and addressing digital wellbeing in the world of VR and spatial computing, we give our exploration of imagination and wellbeing far more complexity and dimension. Let us see how it can play out and transform life as we know and imagine it.

Personal Reflections

1. Why do you think wonder is often associated with childhood? In what ways do exercises in wonder remind you of what you experienced as a young person?
2. How does an intention to use approaches to technology to open up creativity and wonder change the narrative surrounding wellbeing?
3. Thinking about the third facet of wonder, *Surfing Wonder*, when have you ever felt this sort of flow state, and how have you responded to that feeling and cultivated it?
4. Have you ever had an epiphany that seemed to emerge from being in a state of wonder? What about the openness helped to make that possible? Can you imagine how technology as an imaginative vehicle could help support that way of expression?
5. What is an intention you would like to set right now, associated with wonder and digital wellbeing? What's coming up for you, and what are you looking forward to exploring more in the future?

Reflections for Leaders

1. When you think about awe, especially collective effervescence—the awe of shared experiences as a group—what connection could you make to your professional life and your organization's greater sense of collective purpose?
2. How could practicing the first two facets of wonder, *Intentional Attention* and *Wonder Beyond Ourselves,* develop your listening skills and ability to empathize and respond as a leader?
3. Do you have ways as individuals and teams to foster creative flow states of surfing wonder that lead to creative innovation? If yes, what do they look like in your organization? If not, how can you envision creating more avenues for surfing wonder to happen as a regular practice? How could this help your business?
4. How does the Overview Effect relate to your current strategic goals, and is it aligned with your organization's long-term vision and core values?
5. Keeping in mind the wonder facets of *Overview Effect* and *Hidden Wholeness*, what strategic actions could you and your organization take to enhance your impact on and relationship with the environment, fostering a sense of awe, imagination, and purpose?

Extensions for Further Exploration

This chapter provides many opportunities for expansion and further research. One area that is top of mind for me is to research awe and wonder states in groups in immersive spatial environments. There are many qualities of the body and mind, including physiological and emotional benefits, to study and learn more about as we apply findings to digital wellbeing applications that can help animate imagination and improve full life thriving.

4 Wellbeing and the Well of Being Well

In this chapter, we'll explore what it means to look at "being well" as a capacity and think of a "well of wellbeing" as a wonder-inspiring resource that we all can access. Certain truths underpin wellbeing, and identifying them allows us to understand better how to engage with wellbeing every day.

We will also take a broader historical perspective, looking at how several decades of digital transformation have led to this very moment, which is such an important one. We will listen to what young people are saying, using their insights to inform digital wellbeing and how it affects the spaces and workplaces of the future. We can use the wisdom of Gen Z to inform what to institute in personal and professional environments. You'll leave with more resources and suggested approaches and practices, along with an understanding of why intention makes such a difference.

Digital wellbeing, as we have mentioned, involves media that can be medicine or malady. It all depends on how it is created, delivered, and received, and what it evokes in terms of emotions, behaviors, and habitudes. It's about the fullness of the experience itself, what comes before, and what follows after. That is why it takes so much consideration.

We might unwittingly find ourselves drinking from a polluted wellspring when it comes to digital media, and it's time now to pay attention. There's a new well we can build, and it has to do with our commitment to "being well" based on what we consume.

There's a famous expression, "the canary in the coal mine." Starting in 1911, miners used to take canaries down into the mines with them. The canaries have such a rapid, sensitive metabolism and are so oxygen dependent that they would detect impurities in the air and visibly respond to them far more quickly than human miners. So, they were the first indicators of invisible threats. The practice ended in 1986 when the birds were replaced by automated devices.

The younger generations, including Gen Z and Gen A, just might be the "canaries in the coal mine" when it comes to digital wellbeing. All generations can learn a lot from them and adjust lifestyle and expectations accordingly. What they are signaling cannot be ignored. In many cases, as we talked about in the previous chapter, they are telling us that the world is in a state of suffering, loneliness, and hopelessness.

The State of Suffering in Youth Is a Signal for All of Us

Yes, it's been a *fast* fast-forward over the past three decades that involves a generation coming of age as digital tools and interfaces have been emergent, intoxicating, and interconnected. It's also often complex and messy in ways that emotional intelligence has not necessarily been on pace to deal with and meet with equanimity.

In sum, younger generations are suffering. In 2014, the United States saw the largest number in teen self-reported depression that it had on record. There were also more public outcries in responses: anti-bullying campaigns sprung up, support groups were formed, and media coverage of the perils rose. These responses, while they might have lessened the depression trend, did not serve to mitigate it.

The reality bears repeating: a rise in loneliness is one of the biggest epidemics we face as a global society, across all age groups.

As a society in general, we have grown increasingly aware of the importance of wellbeing and the threat of social media. We now understand the threat of too much attention and thereby time placed on social media.

Technology addiction was documented but unsubstantiated until 2019, when it was recognized as an official addiction by the American Psychological Association. Gen Z and Gen A are just beginning to identify their own symptoms and advocate for wellbeing in new and important ways. They ultimately want to build their resilience and thrive, to have resources at their disposal that help them feel energized and whole, as opposed to drained and robotic, responding to apps and algorithms that are short-term focused on maximizing attention over true agency and freedom.

Those of us who work in the digital tech arena also understand the danger of a single story, especially when it comes to the media. Just as we know that technology can use algorithms that promote unhealthy behaviors, we also know that there are ways to approach gaming and immersive technology that are healthy. We know there are ways to forge meaningful social connection, using technology as a bridge. We also know that many of the systems we have created are broken in society. The beauty is that we also have the tools to fix them. The time to do so is right now.

We need to do even more than listen. We need to act and to also give young people agency and empowerment. They are telling us there is a feeling of hopelessness when a small number of people hold most of the power and use that power to perpetuate their own gain at the expense of others. It does not have to be a zero-sum game.

Young people could also be showing us that we have many more options lifestyle-wise than we thought we did and that part of this freedom involves breaking expected patterns and norms. Results could be surprising, inviting us to approach tech in new refreshing ways that can reenergize rather than deplete. There's a lot to embrace, as we stay open.

We will look at these behaviors as possibilities and habitudes rather than prescriptives. It's all open invitation to explore, experiment, and play after all!

The Dawn of a New Digital Era

Party Like It's 1999

As we look to the future, it's helpful to also mark where we have been and how we got here. Readers of a certain age will remember a prophetic mood back in December 1999. There was a distinct heightened anticipation, along with a tinge of panic for some, that rose as "Y2K" loomed on the horizon.

Imagine that you were about to enter an age that many had foretold as the "end of the world," at least as human civilization had known it. It's not hard to imagine; we are in that parallel situation now, with the discourse on AI. In 1999, there were various speculations, including a prediction that all computer systems and technology networks would crash, leaving everyone disconnected and without electronic records. From banking to education to the relatively fledgling e-infrastructure of the social web, everything was projected to be in jeopardy by those who feared the doomsday was upon us. Meanwhile, others took the singer Prince's words to heart and *partied like it's 1999* . . . because it was! It was an unrivaled celebration to mark a new dawn.

Socially, where you were as 1999 turned to 2000 was what mattered. Whether it was a public space, or at home cozy on the couch while watching the Times Square "ball drop," your New Year's choice of location reflected the physical proximate. Where you were determined who you interacted with, as there were very few mobile phones at that time. It can make us nostalgic, right?

There are so many parallels between that analog time and our digital present. Today, one of our greatest "digital dilemmas" is where to store all of the data for our mementos: globally, our pictures and video files take up the majority of our cloud data storage space on average per person. Music and audio files use up a significantly smaller portion of the average person's storage, making up only 6% of overall storage.[1] That said, are we holding on too tightly to nostalgia? What does this do for our mental and emotional health and wellbeing? It's still unexplored territory, and my hunch is that it's having a subtle yet palpable negative effect.

Back in 1999, in the physical local moment, disposable cameras would have been sitting on tables waiting to be used to capture the social scene—this was commonplace and part of the culture of the time. Pictures were printed in labs and at pharmacies, with the option of "singles" or "doubles," which meant that two copies were printed: one to keep in-hand and the second to give away. Maybe doubles was a form of "tagging" in the past! These were the photographic mementos, and whatever happened at the party, aside from those special images and firsthand or secondhand or even *thirdhand* stories, usually stayed exclusively shared by those in the physical localized proximity of the moment of happening. The "now" was salient and rather static in comparison.

Those who were toddlers were being tucked into bed early that night, unaware of the turn of the century. As the anxious masses were reassured by the fact that nothing much happened to disrupt the way of life, a larger shift was about to take place, as these toddlers represented the beginning of a generation who would grow to appreciate interconnectedness in a new way. They would see time, culture, and connection differently, as technology afforded them a chance to grow up with digital, hand-in-hand, as a regular part of their day. So surrounded by devices and interactive immersive tech, one could say they are a generation native to its many worlds of possibilities. In a 2001 article, Marc Prensky coined the phrase *Digital Natives* for those who were "born digital" and are "'native speakers' of the digital language of computers, video games, and the internet."

I believe right now is a key turning point, when digital worlds are becoming spatialized and three dimensional, and they are perceptually *very real*, so the term "virtual" might need to be repurposed to "spatial." Combine that with AI technology and you have a Web3+ future where everything is possible, and the most valuable assets are our data, our intentional attention, and consequently, the quality of our relationships, health, and wellness. It's now a multilayered spatialized AI-powered form of digital, hence the focus on digital wellbeing, a topic that encompasses everything about the way we live.

It's now a multilayered spatialized AI-powered form of digital, hence the focus on digital wellbeing, a topic that encompasses everything about the way we live.

We have traditionally treated the "future of work" and "future of life" as separate concepts, and now those two are starting to be viewed as one and the same. Perhaps in the near future, AI will automate all of our routines and work "tasks" so we will have a unified, integrative "life" focus where we as humans will not have to delineate or label part of life as "work." We will also have less traditional work to accomplish to have a high quality of life. This could be an age of abundance,

Our kindness and care for each other, our quality of relationships, the value of our discernment and decision-making, and our ability to harness imagination, wonder, and awe are all elements that will have the most impact on our quality of life thriving.

powered by play. It used to be that we needed to focus on optimizing productivity. Now, in this period, AI can help us not only increase productivity at scale, it's back to quality of life and the effective sensibility of our care for each other, our relationships, the quality of our discernment and decision-making, and our ability to embrace imagination, wonder, and awe that will have the most impact on our quality of life flourishing. It's back to those new KPI's. Imagine, our basic human needs of food, shelter, and safety could be covered. This is part of why wellbeing continues to be so important.

We need to consider the stakes at every level and every age in society, and we need a special focus on young people, as they are the segment with the greatest decline in reported quality of life satisfaction as well as the most rapid increase in suicides.

The first generation of digital natives are millennials, generally born between 1981 and 1996.[2] Generation Z followed. Born between 1997 and 2012, the oldest Gen Zs were only seven when Facebook launched in 2004. In February of that year, Facebook had 650 active users on the site, and by the end of that year, that number had grown to 1 million users. I declined to join Facebook for more than a decade, citing reasons similar to those causing many digital natives and others to curtail or drop social media.

My article "Facing Facebook" came out in April 2009 in the *Providence Journal* documenting my rejection of the popular site. At the time, all of my friends were on Facebook. Still, I never felt left out. I loved that my photography at the time remained sovereign. I questioned the ethics. When I finally joined, it was reluctantly because I started a business and it was an easy way to share about live events with the general public. I still question and evaluate all of the ethics and data policies underpinning each and every social media platform and app in existence.

My advice in evaluating digital social media for their wellbeing functionality is to pay attention to your own body's cues (for example, are you tired or feeling overwhelmed while using the application; are you more or less energetic after using it compared to beforehand?) and to "follow the incentives." Charlie Munger famously said, "Show me the incentives, and I'll show you the outcome." In other words, you get what you reward for. Traditionally, we as a society have rewarded engagement above all else when it comes to social media. It's past time to change the model. We can trace the past to see how things evolved this way, and we have an opportunity to approach AI differently than our first passes at social media. More on that in later chapters, too.

Gen Z: Growing Up with Digital Media

When Gen Z turned 10, the iPad came out and became one of the most popular gifts for children their age. Even schools began adopting iPad curricula components to try to match the sweep in popularity.

When they were 12, it was 2009, and smartphones had begun to have mass adoption. They asked their parents for ones to honor their teenage years.

When they were 14, a game came out called Fortnite, and they stayed up late playing it, showing up to school sleep deprived and filled with stories of social games and battles the night before.

These Gen Z digital natives are still growing up, now joined by Gen A, and they are ready to take on digital connectivity in new ways and invent new futures. They are optimistic and adamant about ways they want to live and thrive. How do they want to embrace digital wellbeing?

One could say they are "well metaversed" in virtual reality and immersion in general. They might not all have headsets and VR/AR/XR devices, but their brains know what it means to think in spatial, to use avatars, to have nicknames and maps of interaction in 3D. They understand 3D Web in ways that go beyond buzzwords by nature.

They have taught their younger siblings to be digital natives too. It's a group that craves belonging, inclusion, and agency.

We as older generations want to learn from them. And they want to teach us.

Digital Natives and Wellbeing

While they understand so much about digital apps, programs, and games, it's a dire situation for digital natives and younger generations when it comes to wellbeing, and they also hold tools and answers for what is needed.

A rise in loneliness is one of the biggest epidemics we face as a society. Gen Z and Gen A are deeply impacted, and their youth means that they do not necessarily yet have the coping skills to meet the emotional complexity of feelings of loneliness. They need mentoring and support. Studies have shown that teenagers in the United States and worldwide increasingly report feeling lonely, even during a period when their internet use has exploded.[3]

The Social Dilemma, a documentary about social media, focuses on coverage of the ways social apps and algorithms have been built to retain attention and rarely release it. Some have criticized its fear-based tone, saying that it raises awareness and a sense of urgency without suggesting any viable solutions. One of the film's creators and the aforementioned cofounder of the Center for Humane Technology, Tristan Harris, says, "We're training and conditioning a whole new generation of people that when we are uncomfortable or lonely or uncertain or afraid we have a digital pacifier for ourselves that is kind of atrophying our own ability to deal with that."

Young people are telling us all to wake up and that something is wrong with the way we have built things. The statistics speak volumes. There has been a 54% increase in suicides in the 10-to-24 age group between 2007 and 2020. As the *New York Times* reports, "In 2019, 13% of adolescents reported having a major depressive episode, a 60% increase from 2007. . . . Numerous hospital and doctor groups have called this rise in depression a national emergency, citing rising levels of mental illness, a severe shortage of therapists and treatment options, and insufficient research to explain the trend."[4]

Suicide is not just on the rise in young people. It's affecting society as a whole, and the way it impacts younger generations is signaling a widespread problem.

The Stories Underneath Suicide

It's more than numbers and statistics. The data tell stories.

The rise in suicide rates in the United States has been attributed to a variety of factors, including declining mental health, economic stressors, and social changes. The increase in suicide rates has been particularly notable among middle-aged and older adults in the 2000s and then among teens and young adults in the 2010s onward. The Great Recession has been pointed out as a potential contributing factor to the increase in suicides among middle-aged adults, while a rapid increase in depression has been associated with the rise in youth suicide after 2010.[5]

Suicide is now the second leading cause of death among Americans aged 10–34.[6]

The rise in suicide rates across the general population, particularly among young people and military personnel, is multifaceted. It involves mental health challenges, socioeconomic factors, access to lethal means, and unique stressors related to school and military service. Machine learning models, like Long Short-Term Memory (LSTM) networks, have been used to produce accurate estimates of state-specific suicide rates using real-time data, including social media. These models can help in the implementation of timely suicide prevention programs and policies. However, ongoing research is needed to improve these models and reflect the complexity of suicide determinants.[7]

How Can We Use Tech to Prevent Suicide, Heal Loneliness, and Improve Connection?

Dr. Mary Bartlett is one of the world's leading experts on suicide. She has been studying it for three decades and travels globally to speak about the topic. She helped establish the US military's suicide prevention program and now teaches Air Force leadership how to build resilience and be both suicide conscious and proactive in building suicide and wellbeing-related topics into its curriculum, which includes the use of virtual reality tools and other digital media.

Dr. Bartlett frames substantial and complex concepts and ideas in ways that are approachable and storied. When we talked about growing trends that can lead to suicide risks, she emphasized the importance of decoupling loneliness from isolation. Being lonely and experiencing loneliness is not the same as being in isolation, though one can influence the other. She also spoke about belonging and connection as prime motivations and necessities for humans. We're no longer physically separating our work from home, and if we are reachable anywhere, how much time are we giving ourselves away from our work lives?

Dr. Bartlett and I talked about connection and mental health, which can mean different things to different age groups and in different cultures and contexts. We also discussed the vital topic of friendship. "Younger people have 5,000 friends online, but how connected are they really? And then how do we define friendships, right?"

What does it mean to truly connect, and to build resilience through healthy connection? To address some of these questions, I naturally turn to my students, who generally represent Gen Z. Hearing directly from them, in their voices, about healthy relationships with tech, and how they reconnect, is revelatory. "Connection" has a wide range of definitions, some of which serve overall wellbeing, and some which surprisingly counter it, in their views. We'll hear their reflections and participate in our own digital wellbeing self-assessment later in this chapter.

When it comes to building strong relationships and resilience, Dr. Bartlett emphasizes, "It's the eyeball from me to you that makes a difference. There's no replacing eyeball-to-eyeball. So now you've got a generation of people who can break up with each other via text. But really, when you want to build resilience, it is our older generation who taught human skills and conflict resolution and intimacy. Again, it's about face-to-face, eyeball-to-eyeball, I must deal with my emotion in the moment looking at a person in the face. That's related to wellness and resilience."

The eyeball-to-eyeball is also a metaphor. There is a history of evidence supporting small-group therapy and intervention practices that begin with practicing "seeing" someone. While making eye contact may be awkward and unnatural for many groups and cultures, powerful human connection is what we are talking about. There is a human ability for mirror neurons and neurochemicals to sense and respond to each other. Eyeball-to-eyeball, heart-to-heart matters. It's now my own challenge and personal investigation to show how that level of care and connection takes place across digital media. Can those powerful, intimate forms of media, including immersive, provide eyeball connection as a metaphor, if not literally? I would say that they can satisfy both, and I can draw from my direct experiences as case studies. The future can be brighter.

Everything is connected, and connection itself is what matters. The causes leading up to suicide include a range of nuanced topics, factors, and considerations that Dr. Bartlett and I discussed. She emphasizes that both prevention and *postvention* matter. A whole set of programs and treatments can be present for a community in the aftermath of suicide that can help the overall human-to-human relationship in healing, responding, and being together. The postvention strategies operate in a cycle that reinforce prevention methods. It's a positive upward spiral. This whole approach, and our emphasis on eyeball-to-eyeball, heart-to-heart connections are still something society needs to address in a systemic way that includes layers of digital experiences.

The consistent rise in depression, loneliness, and anxiety tracks with the onset of digital tools and media. Correlation is not causation, and I wonder about the ways that digital technology and media continue to saturate our lives. The data-driven links suggest that online environments and digital tools can both pose risks and offer opportunities for supporting mental health. It is important to consider the complex interplay of individual, social, and environmental factors when addressing the rise in suicide rates.

Digital environments can bring people together in many ways and offer that sense of connection and belonging that can be so vital, in ways that are human-centered and authentic. We will talk about that more in following chapters too.

Does a vital sense of connection and belonging need to always come from humans? Are there ways to integrate two forms and have more of a system of belonging that uses the best of what digital technology can bring to support us? We will keep coming back to these questions. I have seen digital methods forge connection quite strongly, serving as a support for humans in ways that do not replace humanity. They are augmentations that liberate.

Suicide prevention using digital wellbeing methods involves a variety of strategies, including mobile health technology interventions, digital suicide prevention tools, and digital therapeutics. These tools include smartphone applications and wearable sensor-driven systems. Technology can help identify high-risk individuals and provide them with the necessary support.[8]

Digital therapeutics can also be used as an adjunct to traditional mental health care. These interventions include automated cognitive behavioral therapy, digital speech analysis, facial emotion analysis, and cloud-based technology.[9]

Studies are showing that AI can be successfully used to help prevent suicide by identifying individuals at risk and providing early intervention—through predictive analysis, natural-language processing (NLP), virtual assistants and chatbots, and training applications to better upskill support personnel. The effectiveness of these interventions in reducing suicide-specific outcomes is under ongoing review, and it's an area showing constant improvement and possibility in arenas of social, emotional, and mental health that have traditionally been understaffed and undersupported.[10]

While AI has the potential to significantly aid in suicide prevention, it is important to use these tools responsibly, ensuring privacy, ethical considerations, and the involvement of trained professionals in the intervention process. I do not think anything can substitute for the heart-to-heart mattering of a human connection.

What Gen Z Is Saying About Goals and a Sense of Hope

We need to listen to stories of struggles and reports of suffering, especially from young people. Young people are telling us there is a lack of hope. And they are also showing us that they need a different approach to wealth and savings that encompasses wellbeing.

The majority of Gen Z are financially conscious and concerned with wealth building, with many exploring nontraditional ways of earning money, such as side jobs and entrepreneurship. As they focus on saving and investing,[11] they face significant financial challenges. More than half (53%) see a high cost of living as a barrier to their financial success.[12] They are grappling with issues such as high inflation, expensive college costs, and a competitive job market. In the face of this, they are cutting back on spending and modifying their lifestyles due to inflation.[13]

In terms of their approach to personal finance, Gen Z is showing a trend toward "soft saving," which prioritizes personal growth and mental wellness over traditional saving methods. "Soft saving" does not mean choosing ease over rigor; it's a financial approach that emphasizes present enjoyment and mental wellbeing over traditional saving and investing for the future. It is a response to the "soft life" trend, which rejects hustle culture and prioritizes self-care and fulfillment. Unlike traditional saving, soft saving focuses on spending for enjoyment in the present rather than aggressively saving for the future. Examples of soft saving include prioritizing experiences, self-care, and mindful spending over material possessions, as well as spending on activities that contribute to mental wellbeing, such as regular exercise and enjoyable work.[14]

Younger generations could also be showing us that we have many more options lifestyle-wise than we thought we did and that part of this freedom involves breaking expected patterns and norms.

We can use their insights to reexamine our own expectations.

What does it mean to prioritize digital wellbeing? How can we see and be moved by the beauty of the world even as we recognize the urgency of the situation? How do we keep ourselves from being sucked into systems that are competitive, jaded, volatile, and untrustworthy? How can we increase and prioritize freedom, joy, belonging, possibility, and love? What do we want to give to the world, and how do we listen to a greater calling to find fulfillment? How can our "tracks" become "pathways" instead, with more opportunities to look out for each other, to look out for the young and the old, to look out for the beauty of this planet?

In an October 2022 op-ed for the *New York Times*, Dr. Jamieson Webster writes, "We seem to have forgotten that adolescents are lightning rods for the zeitgeist. They live at the fault lines of a culture, exposing our weak spots, showing the available array of solutions and insolubilities. They are holding up a mirror for us to see ourselves more clearly."[15]

Can we look in that mirror and face ourselves? Can we be open to necessary changes? There's a lot at stake.

In the next section, we look at habit changes and new ways of approaching digital media that offer more agency and freedom, reflecting our intentionality.

Healthy Habits and Comfort Crushing in the Name of Wellbeing

In terms of wellbeing, most of us, including Gen Z and Gen A, are interested in the benefits of the *here and now*. We do not want to suffer for an illusory future gain.

Writers like James Clear talk about the importance of performing small tasks that are harder in the short term and have long-term benefits. If we understand the benefits of what we are choosing—opting to exercise when it's difficult and challenging because we understand that it's a health benefit that aligns with our goals, for example—we are more likely to keep choosing it and sticking with it.

Routine formation is such a fascinating, nuanced topic because we are creatures of habit and yet we also love surprises and newness. That old adage, "people never change," is being disproven by new scientific findings about the plasticity of our brain and our ability to adapt and form new connections and new meaning, behaviors, and understandings. A lot of it has to do, in my own interpretation, with our identity formation and our level of attachment or openness. This is part of why approaching digital wellbeing with a spirit of openness makes such an impact.

I find that being open to change, and encouragement toward developing self-identity can have a tremendous impact on wellbeing, and digital environments can help by reinforcing new approaches to storytelling about identity and growth. For example, we can use avatars in game design and spatial experiences in meaningful ways to reshape the stories we carry about ourselves. A lot of what I practice and guide involves the reframing and reshaping of stories, where we get to follow a Hero's Journey to a greater understanding as we reintegrate back into the physical world.

In terms of digital wellbeing, we can design our digital environments to support healthier habits, looking at the identity and systems that reinforce our intentions. For example, we can manage digital triggers that prompt our behavior, such as notifications from social media or email, to prevent them from becoming distractions that take us away from work and habits that are more beneficial.[16]

Habit formation and change does not have to be all or nothing, and it can involve small, incremental shifts that are less intimidating and also allow us to somatically sync with our bodies, checking in to nourish our self-care and use our physical sensations to inform how we are feeling.

When it comes to digital wellbeing, this can help us embrace more curiosity and growth rather than reflexively giving into addictive algorithms. Writer Michael Easter has explored the relationship between digital media, our psychology, comfort, and our overall wellbeing.[17] Addressing our addiction to constant brain-stimulation with technology and how it affects our perception of time, Easter suggests that we need to live more in the present and reduce our obsession with technology to improve our mental and physical health.[18]

One way to do just that is to employ what I call "speed bumps" or small wake-up moments that interrupt our pace when we are on digital apps, especially social media. Of course, that runs the risk of interrupting flow state, which we discuss in our investigations of VR as well as animated immersive wonder. It's an area I would love to deeply review as part of ongoing research. Can we have a noninterrupted immersive flow state that is not an indicator of any sort of addiction? My gut and intuition from experience tells me yes, and it has more to do with intrinsic versus extrinsic motivation and intention, plus ways to ease out of the flow state at any time. It's the individual who must have agency at all times.

It's important to keep encouraging ourselves to embrace discomfort, reducing dependence on modern comforts and conveniences, and integrating modern science with evolutionary wisdom for improved health and wellbeing. We can go back to the four culture cornerstones—Dignity, Freedom, Invention, and Agency—and keep using them as our guides. We should not rely on devices or any sort of digital technology for our own feeling of worth and vitality. It should never be a need. We can all encourage less dependence on technology and more presence in the moment.

When we create meaningful media and design games for connection, we look to break addiction cycles and destructive habit formation. We must reject the tactics of the past that did not serve us well. We know that humans are attracted to things that present us with an opportunity to gain rewards;

when the rewards are unpredictable, they become even more enticing, spiking our dopamine. Quick repeatability encourages us to keep coming back for more, so companies profit quickly, and we are overly stimulated. These tactics are used to create addictive loops: in gambling, gaming, and all sorts of activities. There are ways we can break the vicious cycle through raising our own awareness, and also through creators, designers, and consumers changing what they prioritize and endorse.

Changing the Models and Offering More Space for Intention

Speaking of breaking cycles, some of that necessary break comes from establishing those "speed bumps" as part of the ritual, offering breathing space for reflection and for the *intention* to come back to our *attention* practices. One founder changing this landscape through the introduction of a digital tool to help is Royce Branning, who launched an app called Clearspace that addresses tech addiction head on by helping you set limits on your device time. You consciously decide how much time you want to spend on apps, and Clearspace helps you keep that intention. I've used it and found it supports me in keeping to a budget, and I stay much more productive and efficient, without any struggle or guilt.

Branning says, "We believe there's a fundamental misalignment of human attention and intention. And that's a pretty solid ground to build from. If we can move towards greater alignment of what people are intending to do and what they end up doing when it comes to engaging with digital devices, that is a big win." I've gotten to know Branning over several years, as my Stanford classes have used Clearspace in their efforts to develop healthier relationships with technology. Their process involves being more intentional and aware of their choices of app usage and time spent on them. Branning has created an opportunity for more agency right where a lot of us experience the anguish of distraction: our screens themselves.

What if my *intention* is to ride a social media wave of engagement for three hours a day? I asked Branning if Clearspace would "nudge me or judge me" in any way for those choices. His response was, "I think the extreme example of this is if you were to want to watch Netflix for nine hours a day, that's your prerogative. In that case, we would want to help you stay with your plan and wouldn't want you to watch it for, say, nine and a half hours. This puts Clearspace in a position to help encourage you in concert with whatever systems of value you construct around your life. It's up to you. When your discipline is shifting in attention, we want to help you reach that alignment again." I have begun to think of it as an accountability partner. For me, it helps that it is not judging, not human, and easily available and customizable.

It's clear to me that the app can have a positive impact at a personal level, reshaping habit formation and building awareness for what is often invisible and sinister about the addictive nature of social media and all sorts of digital media. I like that it does this in a very fluid, kind, and agency-enriching way. I also asked Branning to address what the broader goal might be at a societal, systemic level. "At our company, we're building the missing intentionality layer of the internet. The 'hair on fire' problem that we have in society right now is causing issues that we can see all around us—relational health compromised, emotional, physical, mental problems. There are some very specific habits that need to be introduced as an intervention in our current state . . .

"As our most optimistic and wild belief, we think that in the next five years, looking at your phone in public is going to be culturally like smoking a cigarette. Forty-five percent of adults in the US were smoking a cigarette every day in the 50s. Now that statistic is 13%. We think that there are trends in society that can put downward pressure on these negative habits. We think that we're in the perfect spot in the next five years to curtail that constant checking of the phone, and all of the implications that has in people's personal lives and in society . . . 2024 is the year where we see that graph of usage start to trend down because we're at that 45% number with American adults today."

> *"In the next five years, looking at your phone in public is going to be culturally like smoking a cigarette."*
>
> *- Royve Branning*

We know that percentage is too high, and Branning and I agree that it's a time to focus on what is empowering.

This period is pivotal. Big tech and creators of social media cannot deny that there is a problem, and the public is better aware and clamoring for solutions. Branning said: "Researchers are publishing an increasing body of work about digital wellbeing. They have been doing that for at least five years, and it's reaching a critical mass. The

Surgeon General has issued a warning on social media for kids, and there are now all of these environmental factors that you can look to as indicators. . . . Now, we can deploy effective solutions that get people going in the right direction, at the speed of software."

Meeting people where they are at and reminding them of intention sounds so elemental and obvious that it's part of the brilliance. In the short term, it's hard to form new habits, and apps like Clearspace are making it easier to match our best intentions with our attention.

Impact for Leaders: Digital Wellbeing Self-assessment

This is a section of the book designed for each of us, as leaders, to consider how we are reshaping the workplace and transforming our personal and professional lives, through a self-assessment we can engage with while listening to reflections of young people. They can help guide us at a time when connection makes a difference, especially as reports of loneliness are on the rise.

Loneliness is a significant global health issue, with an estimated one in four adults experiencing social isolation and 5% to 15% of adolescents feeling lonely.[19] The US Surgeon General has declared loneliness a public health epidemic, highlighting its widespread impact and the need for immediate action.[20] Loneliness increases the risk of cardiovascular disease and has a mortality impact comparable to smoking 15 cigarettes a day.[21]

Even though hope is at stake and the state of global health is dire, there is also an optimistic and practical quality to younger generations' approaches to taking action to support their wellbeing. They can be our guides.

Mental health issues are becoming destigmatized, and digital natives are setting the standard for other generations by openly talking about struggles with emotional and mental health. Their wisdom can guide our leadership practices, transforming how we approach our professional and personal lives too.

I design and teach a course at Stanford University called *Digital Wellbeing by Design: Healthy Relationships with Technology*. The title is a double meaning: we examine our relationships to technology itself, as well as our human-to-human relationships mediated by technology. Students are drawn to the topics for various reasons, often motivated to take agency and play the role of a designer. They want to improve their own approach to tech and health.

They're motivated by all sorts of factors. Some want to invent wellbeing apps and programs that are wildly successful at scale, some want to become healthcare professionals who use digital wellbeing in their approach to medicine, some want to change behavior patterns and lifestyle in relationship to technology use, and others are creative and want to focus on expression and the arts related to tech. Some are simply there for an animated conversation and open learning—which does not disappoint. It's digital wellbeing by design.

Regardless of what brings them to the course, what keeps them involved in the dialogue involves *finding themselves* there. We cannot be abstract. We cannot be all philosophy without the investment of what is personal in practice. We cannot extrapolate any sort of larger meaning without the matter of what is personal. And we cannot do anything without the authentic connection, conversation, and the "eyeball-to-eyeball" moments. It's about human relationships, after all.

Many of my students share they are excited by the promise of tech for wellbeing, and the majority will also share they feel some level of stress and depletion. They're high achievers by nature, and they feel pressure to perform, and to be seen as successful. This pursuit is often a futile endeavor. When it comes to success, what is enough?

The general topic of wellbeing is fraught with its own natural stumbling blocks, discomforts, awkwardness, and anxieties. Mindfulness practices are introduced throughout the study of digital wellbeing. Everyone is encouraged to treat their own bodies with care and dignity as they are studying wellbeing in general. This points to what it actually means to live out the messages. We cannot approach a study of wellbeing without aiming to *be well* ourselves. Or we *could*, but it would likely be short-lived and a rather hollow endeavor at that. It's a state of being.

As students approach the design of a relationship with tech from all angles, many have said that it becomes nondual, not a "yes/no" approach to tech, but rather a design opportunity, where they have more proactive strategies to incorporate as they look to match their own intention with attention. It's digital wellbeing by design.

We talk about how digital wellbeing through a lens of wonder can essentially become mindfulness in practice, within the context of a digitally mediated world. It's about the media we encounter and consume on a daily basis. As we will see in coming chapters, the media is all around us, a pervasive part of our lives, and there are ways to be more mindful in the ways that we approach it, and for media itself to become more mindful and medicinal in the way it's designed, created, and delivered.

As I was in the process of writing this book, I openly asked students in my course to participate by weighing in on their current feelings surrounding technology, relationships, and the world in general. They gave permission to directly quote their answers. The best way for us to learn from them is to listen and reflect. We can use these questions and answers to illuminate our own leadership practices and goals.

The diverse academic interests of students in my Digital Wellbeing course underscore the universal relevance of the subject. With ages ranging from 19 to 26, these students—majoring in fields as varied as philosophy, biology, art, technology and society, and mechanical engineering—reflect a microcosm of the university's intellectual landscape. This convergence of disciplines and diverse perspectives within the course illustrate a shared recognition of the critical role digital wellbeing plays across all facets of academic and personal life.

When I asked the AI tool "ChatGPT" to evaluate and summarize students' answers, painting an overall picture from each data-rich answer set, I found results far less exciting and electrifying than students' pure answers. It's as if AI had used a blur tool on the data. This is one area that highlights how AI can help make a process more efficient, and it also can obscure authentic voice.

For this exercise, I invite you to **first consider the self-assessment from your own perspective, *before* reading anyone else's answers. Think about your own workplace, especially, and your colleagues and collaborators.** Consider your corporate structure, and perhaps use your own primary answers to inform ideas about how it could be changed for the benefit of wellbeing. Include your personal life by holistically addressing the questions. Then, you can juxtapose your ideas with those of the students. This is one of my favorite exercises to do too—to self-reflect and listen to student voices, letting their stories about struggles, goals, passions, and joy inform and sharpen my awareness. You'll see their honesty, vulnerability, openness, and hope come alive in their words. Experiencing this exchange is an exercise in wonder and awe. We can use it to transform our approach to work and life.

Digital Wellbeing Self-assessment: Thriving with Tech

The full assessment, along with the full range of student answers, can be found online at caitlinkrause.com

Why Did You Look to Approach Digital Wellbeing Through a Lens of Imagination? (*What Brought You Here?*)

"I want to investigate how I can use tech to enhance my life, not just simplify it."

"We can discuss how various facets of technology (wearables, VR, social media, etc.) inform our understanding of wellness, and explore strategies to optimize wellbeing in an increasingly tech-oriented world."

"The course made me reflect on my own thoughts and actions often, and in doing that I was able to notice positive change and development in mindset."

What's the First Word That Comes to Mind When You Think About Your Relationship with Digital Technology?

"Symbiotic"
"Overwhelming"
"Addicted"
"Complicated"
"Connected"

Why Do You Think That Word Came to Mind?

"I need technology to do my best work, and in turn, I help improve technology in a manner. Although, symbiosis implies that there is no harm being done to me (or the computer), which is not always the case."

"Recently, it has felt like I have had to learn more self-regulation and self-discipline when it comes to my tech use. It feels like an obstacle toward deep connection to the world that surrounds me in the present moment."

"I think my life and largely my interactions are heavily reliant on digital technology. Even for my potential career, it will nearly be inseparable from digital technology, though it is grounded in the real world."

What Digital Tool Do You Use That Brings You a Lift? (i.e., You Start Using It, and You Feel Overall Better After?)

"Siri to set up a timer; sleeping app; google calendar my Notes app. I love typing my thoughts there."

"Music apps on my phone are a good way to stay off the screen and enhance my experience while I do daily tasks."

Have You Ever Experienced a Feeling of Being Creatively "in Flow" in a Way That Sparked Imagination? What Brought That About?

"Yes I have and it was when I was by myself. I did not have any distractions such as TV or music that was playing while I was doing work. I was in the library in a section that was pretty bright but quiet. I was definitely in a flow in terms of being locked in and getting my work done."

"I am in the flow when I am not too hungry or too full, my core body temperature is just right, there are not uncomfortable sensations in and around my body, and I am listening to the music that best matches my current mood."

"I think it starts with an attention to the ambience of the space, making sure there's enough light in the room—sunlight if it's day, or a cozy warm light if it's night—making sure my space is clean, setting instrumental music at an appropriate volume. It helps start the state of flow, but once I'm in it, they become almost invisible."

Do You Use Technology in Any Way to Help Support Positive Habits? If So, What Do You Use, and How Does It Affect Your Lifestyle?

"Yes, I use technology to keep relations with people that I may not get to see very often. This includes my parents, my siblings, and my friends. It allows me to check in on them and make sure everything is going well for them while also making our relationship stronger so I have people to talk to in times of struggle."

"I use my phone as an alarm clock to wake me up in the morning, and I use my kindle at night so I can read in the dark . . . I use the clock to time breaks, and it helps me get into the flow . . . I take my notes on my iPad using virtual paper (Notability) as I do not want the extra weight of paper."

How Would You Describe a Healthy Relationship with Technology? What Are the Key Ingredients?

"The key ingredient to a healthy relationship to technology is agency."

"A healthy relationship with technology essentially to me means having technology that enhances your day-to-day life. Enhancement can come in many different ways—I'll typically use technology during free time when I'm bored to prevent myself from simply doing nothing, sometimes doing nothing at all can become bad for me and sometimes even saddening."

"I should determine what I want. Sometimes, apps and games give me a sudden notification to attract me to do something, but I do not like that kind of thing. I want to use tech only when I want it consciously."

"I think discipline is the key ingredient to a healthy relationship with technology (and toward a healthy relationship with life in general). Just like knowing when to stop is helpful, knowing which technology applications are the ones to use for your current problem is another invaluable skill."

"I believe a healthy relationship with technology exists when people are able to access tech in a positive way that does not lead to screen addiction while first being at peace in reality."

"Peace, meaning, connection, positive energy."

"I think it is having a clear intention on the use of the technology before interacting and sticking with that plan once you are inside."

"A healthy relationship in my eyes would be to use it for fun but also being able to put it away or down whenever without having the feeling of missing out on something or needing to go back to it right away."

How Has the Advent of AI Tools into the Mainstream Made Your Life Easier, Harder, or the Same?

"It is easier to create content, or brainstorm ideas, summarize information or provide guidance for tasks. It has also become a crutch for my laziness, which is a current priority of mine to fix."

"ChatGPT is helping me proofreading all kinds of writing. . . . On the other hand, I do not learn from it much. For example, if I used English dictionary to find words or search on Google for codes that I want to use, I could have learned something from there, and I would not need help in the future."

"I have only really used it when coding in Python for classes that have recommended its use. I use Brilliant, and will sometimes use it to help with mathematic and coding concepts that I am struggling to grasp. I do somewhat fear how I will work with it or if it may replace the skills I am currently learning."

"It has made my life harder when it comes to technology because the algorithms are getting better at understanding my interests so that feeling of always wanting more continues to be more present."

What Is Most Needed, Specifically, to Combat the Loneliness Epidemic? Can Digital Technology Help?

"I would love a social app in VR where I could hug my friend and feel the hug through sensation manipulation."

"Connection is an obvious "solution" to loneliness, and I do think technology can help in that respect. During Covid, I was part of a small group of online friends that connected through a game, and they certainly helped push back any feelings of loneliness that I had. This does bring up the question of whether or not online friends are real friends (I would argue that they are)."

"Digital technology certainly helps people contact each other, but does not offer the same connection that being together in real life brings. In-person relationships, I think, are essential to combating loneliness."

"Compassion and internal understanding. Technology can help as it broadens the range/reach for people to communicate with others and experience different environments."

These insights reveal a complex picture of future digital wellbeing, where innovation, personal responsibility, and societal shifts all play pivotal roles. It's necessary to engage in deeper self-reflection and listen to what young people are saying because they can help guide us to institute changes to the workplace and changes to how we address organizational culture. We can influence personal and professional habits and opportunities that will transform quality of life flourishing. We are opening up in support of whole-life thriving.

With every one of these questions, I frame them as leadership guides, as we can first ask ourselves about these universal questions, and then compare them to what young people say, learning from them.

As we synthesize and respond to the array of direct, primary source insights collected here in this chapter, we have been provided with a comprehensive overview that is as instructive as it is enlightening.

The reflections, while varied in scope and perspective, converge on the points that underpin many of the core themes of digital wellbeing. When I consider what young people have shared here, I am motivated to respond through my own life design and through providing many opportunities for digital wellbeing through a lens of wonder to keep offering pathways to self-realization, transformation, and better collaboration and compassion. How have you responded in this dialogue? More reflection questions will close this chapter.

In Conclusion: The Sheer Reality of Digital Wellbeing: Living with Technology in a Spatial World

Spatial computing is here to stay, so perhaps digital wellbeing will be "spatial wellbeing," which circles back to **what a human is in the first place: an embodied spatial creature**. Spatial is special for many reasons, and it's not a near estimate of physical or any attempt to mimic. It is exactly what it says it is: spatial computing. The "virtual" in virtual reality implies approximation, which is not entirely what it's about.

Digital wellbeing will be "spatial wellbeing," which circles back to ***what a human is in the first place: an embodied spatial creature.***

While many people were physically separated due to the pandemic, the ability to virtually be together brought a feeling of real connection. XR and spatial computing make the experiences shared across virtual media very *real*. What is not physical does not mean it lacks substance.

The largest pressing factor these days is the focus on focus itself, the call for attention, which is one of our greatest resources, along with time. What does it mean to attend to and be with someone, and to be truly present with them?

Listening is a skill that leaders often identify as the top distinguishing trait in their field, the "X factor." Most young people admit needing to practice listening (as all generations do!), to tune their awareness to the present moment and filter out distractions. The human brain gets overloaded by new information without context, and we need to take time and attention to really understand story and meaning, which creates the context we need. It's a world with a high noise-to-signal ratio, and we need better decision-making and discernment.

How do humans make **meaning and context** out of all of the data, stories, and resources that surround us? This is what will distinguish us from AIs in the future. It's semantics over syntax. We must apply meaning, and hopefully it will be deeper than a popularity count or an influencer rating.

Generations of young people are longing for meaning to matter, and their meaning-making depends upon all of us helping to reanimate the system, looking through lenses of wonder instead of cynicism. We need our imaginations to bloom as we appreciate the details of experiences and forge connection. We can reframe and transform our corporate systems to reflect new values and visions. Then it's a reciprocal lesson, as young people also mentor and teach older generations about new ways to connect and collaborate in the world. We can all be master storytellers and creators, and we need to start encouraging and listening to each other. That is one of the key ways that we can all draw from, and also refill, the well of wellbeing.

In the remaining chapters of the book, we will explore more of the "how" of digital wellbeing and show examples from a wide array of digital media as facets for consideration.

Personal Reflections

1. Name one person, the first who comes to mind, who has made a difference in your life. Describe how this person whom you admire has made a difference. This can be a gratitude practice and a reflection on friendship.
2. Describe a time in your life when you were among other people at a gathering that was completely "analog"—no posts on digital social apps; no record-keeping; no texting or sharing in any way other than being physically present. What do you remember about all of the details of that place, time, and experience? How did it feel to connect in that way?
3. What "eyeball-to-eyeball" moments have impacted your life? What do you think led to them being so powerful? How could you create more conditions for those moments to happen, for yourself and for others? Can you imagine ways that digital connections could lead to even more of those moments?
4. How do you view isolation and loneliness? Are they two separate topics? Have you ever felt one or the other, and what did you do when you were feeling that way? What digital tools serve to help you?
5. Describe the first memory you have of using a "smartphone." How did your interactions with it feel different than using a traditional mobile phone? Which smartphone app was one of your first to use? How did you come across it? Did anyone teach you how to use it, and nowadays, what apps are most useful to you, and which ones do you consider distracting or unnecessary? Would you like to set a new intention for smartphone use?

Reflections for Leaders

1. Looking back at the "Impact for Leaders: Digital Wellbeing Self-assessment" section, how did you find the process of engaging in self-reflection at first, addressing your work and leadership practices? How did it feel for you to listen internally, and then how did it feel to read and experience the student responses?
2. One question asks what digital tool(s) bring you a lift. What is your response to that question, and how often do you use those tools or methods in your current workplace and role?
3. You were asked to identify the key ingredients to a healthy relationship with technology. Do you see this as a healthy relationship with the tech itself, or healthy human-to-human relationships mediated by the tech, or both? How do your responses, and those of the students, match your current workplace dynamics? Is it currently set up to encourage healthy relationships with tech? How would you like to empower that to improve?
4. In terms of your work, how has the advent or introduction of AI tools into the mainstream made your life easier, harder, or the same? How did your answers to the question match students' responses, and was anything surprising? What are your future goals and strategies involving AI in the workplace, and what intentions do you have about incorporating AI into your workflow?
5. What other strategic implications for your organization were uncovered through this whole exercise and through listening to the ideas of younger people? What might have changed about your views through this chapter?

Extensions for Future Exploration

This chapter's research and conversations reflect my core. In many ways, this entire book is a wonder-inspired response to the voices of young people. I have a sense of hope for the future. Building on this, my intention is to research and share more about heart-to-heart moments, connection, and collective effervescence across digital media. I want to share more about experiences young people are illuminating as well as learn from other groups about how we communicate distress and need for support in many areas related to wellbeing. I would like to deepen the dialogue surrounding how lenses of wonder and imagination can help reframe the future.

I am motivated to expand on the theme of resilience and explore what it means to be resilient overall versus digitally resilient. We can go deeper.

This chapter addresses a range of critical topics related to wellbeing and lifestyle. As additional extensions, I would like to develop the idea of sabbaticals as part of a work routine and incorporate details related to different generations' experiences with stress and burnout. There is a growing trend in modern work lifestyle that involves "sabbatical sessions," allowing time to readdress life purpose and meaning. Investigating this area combined with wellbeing and wonder is one path I would like to follow. There's a lot to expand upon, as I would also like to broaden the conversation surrounding legacy planning, death, dying, and the grief process.

In the near future we will have more robust studies and data pointing to how effective AI is at helping us address and mitigate the loneliness epidemic. How will it serve us as leaders? There are many ways to grow in response to these topics. Naming these extensions feels like stargazing and recognizing constellations. Our North Star is wonder-infused wellbeing.

5 How Technology Helps: Creating Connection and Belonging Instead of Loneliness

This chapter about relationships is a shining bead on the necklace of digital wellbeing because humans are essentially social animals. Whether we are introverts or extroverts, including everything in between, our survival and thriving is interdependent, and we build bonds and new understandings through relational trust. We are living in a digital age where technology mediates the connection and can help to build relationships in ways that are healthy, imaginative, intellectually stimulating, and emotionally nourishing. How can we use better knowledge about AI-informed technology to empower connection rather than disconnection?

At South by Southwest (SXSW) in 2023, Esther Perel took the stage and spoke about the delicate and nuanced nature of attending to relationships.

She coined a new phrase for AI: *artificial intimacy*. Replacing human connection with AI bots is tantalizing because of that "always available, always on, no bias" lure but would lead to the same negative outcome that people suffered when the attraction of chemically processed "always available" shelf-stable food caused the food industry to substitute it for more nutrient-rich choices.

"We've spent the last decade shifting our lives into their virtual expressions," Perel said. "Being uncomfortable, doing things that you end up not enjoying, being afraid and taking risks are some of the ways that we learn who we are and who we are not. The experimentations and the failures are essential to the development of our identity."

What Perel says captures the "comfort crushing" philosophy we discussed in the last chapter. We do not want to live anesthetized and comfort-focused as humans; we crave healthy challenges and rigor for growth. That's what "flow state" is all about, which we'll discuss in a later chapter—it's a balance and an inner knowing that meets the outer world. Some would call that "challenge/skills balance." The point here is that it's not about being fed a continuous stream of our likes and happiness. Life also has shadows and complexity, and that creativity and uncertainty of those relationships are what allow and encourage our development.

That said, AI and complex digital systems can also support these nuances, if we are intentional. Technology does not have to create divisions between humans. It could help to create more bridges.

Perel said, "While predictive technologies have solved many of life's biggest inconveniences, they are also making us unprepared and unable to tolerate the inevitable unpredictabilities of human nature, love, and life."

Digital tools with recommendations and algorithms tuned to reduce friction and smooth rough edges are not the optimal stage. This is a passing phase in the development of interactive digital media, and the evolution of that media, already underway, is much less flat, less 2D, and less banal. My findings lead me to explore the wonders of AI-enhanced digital media that serve to help connect humans internally and to each other rather than disconnect them from themselves and the world, as Perel and others warn about.

It's our own human agency, our freedom to act and witness results of our actions, filled with uncertainty and delight, that remind us of the wonder of human experience. By witnessing our own experiences and those of others taking effect in the here and now, we have a deep sense of presence and belonging. We can tune our senses, too, to be more responsive and curious agents in the world, not passive passersby.

Relationships take work. They are hard and challenging and so often rewarding. But the frustration to get there is real. It's also the friction that contributes to a quality, authentic, healthy relationship. A relationship with a human, as so many of us know, including my students quoted in the last chapter, cannot be replaced by AI.

Why would we even wish for AI to replace a human relationship? I would argue that it's not AI getting in our way. It's not a screen problem or an AI problem. It's not easy to provide a simple blanket solution because we have to look at the complexity of the situation. In other words, a screen or an app or a VR device could be a layer augmenting a human experience, and that experience could involve connecting with another human. When humans distract themselves from engaging with other humans, it's a design flaw in the way that they are engaging. Once we reveal what's actually happening, it becomes less about choosing a screen over someone else.

When humans distract themselves from engaging with other humans, it's a design flaw in the way that they are engaging.

For example, how often does someone say something to you and you need them to repeat it? *What? Can you say that again?* It's not that we cannot hear each other. It's that we are lost in our own thoughts most of the time, either planning what we want to say next or thinking about something else . . . and a lot of people report that what they are thinking about is not a reverie or an imaginative ideation about possibility—it's actually a fear or a rumination. We're often lost in anxiety.

When we are not lost in this type of rumination, we might be elsewhere: on our phones, scrolling, and completely distracted by what can be mind-numbing and compulsive. We could also use those same phones to deeply engage with another human—in a phone call, sharing a video chat, or in a game. Our smartphones are digital multiuse devices, so they can also provide us with useful information, resources, research, and entertainment. The smartphone is not an ergonomic means of communication. The sound is often low quality, the connection can cut in and out, and the screen is so small and two dimensional that you do not see very much and certainly do not usually feel you are transported. What these live calls have going for them, however, is **shared time**. You are sharing time with whomever is on the other end. Forget social media for a moment because I'm not talking about livestreams or scrolling through posts. I'm talking about shared intentional copresence.

A live call is powerful, even if only in 2D, because of that shared presence and attention. And it relies upon digital media. Digital is the medium, and the point is connection and presence. What we attend to is what we care about, and some would call that a form of love.

What is happening in our 3D worlds, when we are together with those whom we care about, whether in the workplace or in our human relationships? How many of us spend much of our working days on a computer and then finish the working day and find we have no energy for anything except passive media consumption? Do we scroll on phones mindlessly . . . and is there ever a person next to us doing the same thing?

This is a cultural behavior pattern that's absolutely concerning. And we have all been there. Imagine we are sharing a vulnerable and important personal thought with our close friend or partner, and we hear a hollow "uh-huh" in response. We know the person is only half listening, distracted by their phone. We have to be patient. There's a gap of time, and we know that they are not fully present, not fully with us. This is artificial intimacy revealing itself.

Enough is enough. We do not have to accept this type of communication as the norm, some zombie-like exchange with the people we care about, vying for attention with each other. This epidemic-sized problem is a socially accepted and culturally normalized distancing from each other through disrupted conversation is something we can effectively combat and surmount. **How do we do this? Through awareness of the power of awareness.**

The problem is not the technology, though. The problem has nothing to do with things like gaming, virtual and extended reality, and spatial computing. The problem is not even about digital 2D tech. The problem is human hubris, in thinking that we are clever and advanced enough to outsmart the algorithms and be able to carry our phones around with us and be ambiently copresent with everyone everywhere all at once. We want it all.

I've been there, too, trying to volley my awareness. We think we can pay attention to everything when the truth is we can only be focused on one thing at a time. When we think we are effectively multitasking, our brains are rapidly toggling back and forth, back and forth. It's not as efficient. And if we in fact are doing multiple things at once, it's much more taxing, and we are not giving our full presence to where we are. So those of us on smartphones who think we are also paying full attention to our friends and partners and children, think again.

That said, our brain does not always have to be focused on what's physically present in the world around us—there's a whole inner imagination-rich landscape to explore and enjoy! I'm a proponent of wander and wonder engagement for work and play. There is ample space for reverie and mind-wandering, and it animates creativity. My previous book, *Designing Wonder,* describes those five facets of wonder, which we also explored in Chapter 3, to explain some of the background and development in this highly powerful field. All of the wonder, awe, and perspective taking would not be possible if the best mindstate were a forced, fixed attentiveness to our daily tasks and our physical environment. There are meta states, surfing wonder, hypnogogia, the overview effect, and more to use as methods and conduits to reach that higher plane of wellbeing. Recently, I talked with Jon Kabat-Zinn about these creative forms of mind wandering and surfing wonder. They are effectively meditation, and we agreed that they are absolutely vital. We can all learn how to access wonder.

Let us circle back to the topic of smartphones and their incessant allure. Yes, we are truly living in an awkward time to be human: the age of the dominance of digital 2D and a transition to spatial computing. Is that 2D age possibly decreasing? I think yes, because it's giving way to an age of 3D spatialized technology, which is on the rise. What we feel and sense now is in transition. Change is awkward and uncomfortable too, so there's a natural resistance and volatility about this time as well.

Maybe you are feeling it even now, as a reader. You might be thinking, "Why is this important? What does spatial computing and 3D tech even mean? How can wonder and imaginative mind-wandering be positive, while distraction on phones and social media is negative? My work team meets in-person and/or sometimes remotely and hybrid, and I know there are career shifts, but how does this affect me right now? Isn't the twenty-first century all about technology and digital change, and our number one goal is to somehow just stay sane in the process?"

I hear you, and from my perspective, as someone who specializes in well-being, imagination, human collaboration, and technology of all sorts—this is an awkward and uncomfortable time, absolutely, and it's imperative that we each learn how to unlearn some lessons from the past—about worth, about productivity, and about imagination and creativity, to name a few. We address all of these questions you might be holding, as well as the paradoxes of this time, through the journey of this book.

The digital technology that can help animate our wellbeing is not only worth learning about here in this text; it's going to be reshaping our futures, so the time right now is the critical one to set foundations with our best values and optimized lives, filled with quality relationships and true thriving. We have not done that in the past analog and digital revolutions, and there's an opportunity now to change that trend.

Digital technology is not the enemy. AI is not anathema. Answers about how to approach wellbeing in an age of spatialized, AI-powered tech are even deeper than integration or holding opposites when it comes to use. It comes back to our experience and our awareness. We're invited here to explore wholeness through taking back agency, and that starts with us and setting our intentions, then matching that to our attention.

In February 2011, Sherry Turkle gave a powerful TEDx Talk called "Alone Together" that captured widespread media and public interest. She delivered a mainstage TED Talk called "Connected, but Alone," along the same theme of human disconnection, in 2012. If you have not seen that mainstage talk, I recommend watching it.[1]

One of the takeaways is that humans need to pay better attention to each other, much like Perel shared. Turkle noted that people could be presumably together physically, yet still ignoring each other while their attention is placed on devices.

It does not have to be this way, and the technology is not to blame. There are many ways to reframe and repurpose our engagement methods with technology, with each other, and with the beauty and splendor of our world. We can take back our agency and match our intention with our attention. We can also set new priorities, learning more about how wonder can transform our ways of thinking about technology in layers. We hold the power of choice.

Have you ever had an experience that reminded you that right now is an awkward time for humanity? I realized it recently while riding on a Swiss train, traveling from Lucerne to Zürich.

In the quietude and peace of that train, what was absolutely striking to me was noticing on this particular ride that *every single person on the train was focused intently on their smartphones*. Whether standing or sitting, everyone was looking at phones. The posture was universal: bent over, hunched, necks straining, shoulders sagging. It was a terrible sight to witness, with the beauty outside speeding past us. I know it's the typical human body posture of "being on a smartphone," and yet it also made me think of the posture of sadness. Meanwhile, here we were, passing through gorgeous landscapes, skirting Lake Zürich on the way to the city, and no one was paying attention to the natural surroundings.

Suddenly, the silence became very noisy, as I imagined all of the conversations, apps, colorful photos, and posts vying for attention. I watched people's demeanors and the looks on their faces. Not many seemed happy in their expressions. What does it say when a shared natural physical experience of riding on a train does not bring some collective joy?

Collective effervescence is one of my favorite aspects of awe, which we also talked about in Chapter 3. Why do I like it so much? Because it is shared. People are moved together by an experience as a whole. We learn togetherness in that essential way that is not a lesson as much as a remembering that we are social animals who exist as a group. We depend upon it for survival, this ability to flock, cosense, and form a collective entity through our social and even physiological synchrony. Our individual perceived separateness can become something we overcome together. It's a network. Biology already teaches us this and programs us for it in lessons on resilience. Even something simple like watching a play together as a group in a dimmed theater, lights on the stage, hearing ambient rustles, cheers, laughter, and feeling the focus together brings this feeling about. Riding a train used to be like that, witnessing and participating in something wondrous together, a shared ritual. That day, humans hunched over smartphones, we were bound for the same destination, but there was no collective effervescence, no awe in the journey.

It's not all about the devices stripping us of power and awe and wonder, though. That's the danger of a single story. Digital wellbeing by design, using a lens of wonder, means that the technology has the power to relay this experience of connectedness and transcendence in many ways, and it's a powerful medium for creativity, wonder, and delight. It's all about the design and the intentionality when it comes to how it addresses and guides our attention, our engagement, and all of the "Seven ThEmes" we talk about with spatial computing. We discuss this more in the coming chapters about spatial extended reality and gaming.

Reframing Through Awe: Turning Midlife Crisis into Chrysalis

As we talk about humans experiencing awe together and question how to intentionally use devices and technology that are truly in support of our human joy, wonder, and belonging, it's a perfect time to reference Chip Conley, an entrepreneur and thought leader who is revolutionizing approaches to midlife. Conley is recognized for his innovative contributions to the hospitality industry and his thought leadership on aging and modern work culture.

Conley's former role as Airbnb's Head of Global Hospitality and Strategy marked his disruption of the hospitality industry. In January 2018, he cofounded the Modern Elder Academy (MEA), the world's first "midlife wisdom school." MEA is dedicated to helping individuals navigate midlife transitions and to reframe aging as a period of growth and opportunity. The academy has a global reach, with more than 3,000 alumni from 42 countries and 26 regional chapters.[2] I can count myself as one of those alums!

Through rebranding the concept of "a midlife crisis" into "midlife chrysalis," Conley's contributions to the world span the realms of entrepreneurship, hospitality, education, and social activism. He is the ideal person to ask about social connection, belonging, and what really matters in this age of profound loneliness and disconnection.

Our dialogue combines philosophy, practicality, and pedagogy, for the short and long view. I started by asking, "Will or how will digital technology change and support how we approach aging and thriving? How could it expand our imagination and our connection, and what should we keep in consideration when it comes to digital thriving?"

Conley answered, "When you spend time with the investors in the aging space (many of whom hang out in the Aging 2.0 organizational ecosystem), they tend to focus on tech products and gadgets that will help the elderly. They're less focused on the social wellness of those in midlife and later. I think technology could be harnessed to create more pods/cohorts of people who have common interests or where mutual mentorship could occur—where one person knows how to bake a great pie but wants to learn all the tools on their iPhone and another person is the vice versa so you can match them. Or match older people who have similar athletic goals so they can exercise together."

I agree—and that matchmaking and bringing people together, using tech as a bridge for those social cohorts, is what it's all about. That's why in virtual reality and spatial computing, one of the areas I like to build is centered on the notion of the "campfire." How can we use that campfire metaphor to bring people together, to share and understand each other more deeply, and also have fun in the process?! There's an excitement happening in the middle of the circle. Spaces in the metaverse can feel like this sort of centered, whole sharing, and those spaces are exploratory as well.

I also asked Conley about the digital tools and media he uses and how they make life easier. He talked about his sleep score machine. Sleep is an important topic as we age, and he described how he experiments with what does and does not work when it comes to his sleep. Practical tools, yes! We're going to talk later on in this book about digital wellbeing and wearables, and sleep is so important. It comes up over and over as a top priority, affecting all aspects of health and wellbeing.

I wanted to learn more from Conley about the way MEA brings people together and forms movement. Focusing on his calling midlife a chrysalis, not a crisis, I asked what he thinks is our greatest asset as we strive to honor long-life learning and open, curious mindsets.

Conley brought up awe and wonder and how they have informed two pieces of the magic of MEA: "The top two pathways to awe globally are witnessing moral beauty and experiencing collective effervescence and I think MEA offers the habitat that allows for that." He went on to address what MEA does to focus on healthy aging, the three main things being "creating social wellness by helping people to feel seen and connected . . . helping people move from a fixed to a growth mindset so they feel comfortable becoming a beginner again . . . and reframing people's viewpoint on aging" based on research that is being done.

These answers not only reach my core, they reflect the driving basis of my work to empower connection. There are three parts: (1) inviting people to **experience wellbeing firsthand** in the way we connect outward-in and inward-out, reframe, and "fill our well" again with reflection and care, compassion, resilience, and curiosity; (2) **imagination** in the way our curiosity engages with the world, propelled by wonder and awe, and that spirit of collective effervescence; and (3) **collaboration**, which happens in a powerful way when there is relational trust, deep listening, growth mindset, and a purposeful sense of collective presence. I've found that virtual worlds can bring out opportunities to practice all three if guided wisely.

When our conversation turned to what makes for a "meaningful life" and how, or if, it relates to digital wellbeing, Conley said, "I have to say that a meaningful life is definitely a connected life, but I would also say it's a life based upon Erik Erikson's maxim 'I am what survives me.' It's about feeling like you have made a difference in the lives of others."

Conley's words here about "a meaningful life is a connected life" coupled with his thoughts on legacy has me thinking of the beauty of a distributed network and its ability to share efficiently and vibrantly between nodes, like a neural network in the brain. This could be a visual framework for meaningful digital wellbeing—to keep using models to share like that, in ways where the legacy spreading involves moral beauty and becomes a thing of awe and oneness.

As we turned to AI and what it portends for the work in the coming years, Conley's insight was powerful. "In an increasingly AI-reliant society in which knowledge becomes a commodity, I think wisdom will be an asset or quality that we seek out more readily."

In his own inimitable style, Conley has left us with more seeds to plant, more wisdom to discover. That our reliance of AI is leading to knowledge commodification, perhaps without the "wisdom journey" of traveling the longer path to arrive at that knowledge, giving it context and color, is not lost on me. It's a concern, which is not to say AI is to be shunned, but the future will be determined by our ability to slow down and consider possibilities beyond the fastest, simplest answers. As Conley says, "Wisdom will be an asset or quality we will seek out more readily." And we will likely need to look for places and (human) guides to help us develop that wisdom, while forming meaningful relationships and belonging with each other. It comes full circle, doesn't it?

The Wonder of Copresence: Shared Connection Across Physical Divides

Something to consider is ambient copresence, which I believe can meld well with collective effervescence. Designer Maggie Appleton calls ambient copresence "a subtle sense of shared, synchronous space among multiple people on the web."[3] I appreciate and intentionally design experiences that emphasize the thrill of being together, which does not always have to involve loud sharing that we ambiverts (those who are both introverts and extroverts) can sometimes feel disturbs the flow.

Imagine the emotional connection possible through copresence in augmented spatial forms. For example, you could go to a physical meditation location on a map and discover an object there layered in a spatial environment that can anchor and represent an emotional exchange. You could tag some thoughts there for the next person to discover. While you are there in physicality, you could discover others around the world, in similar meditative locations close to their physical proximity, who are all focused with the same intention and attention. The space itself could glow more brightly, signaling this. Haptics could trigger. With spatial, everything and anything is possible, and it can be deeply uplifting.

The key factors are the intention and attention placed on what becomes a shared ritual. Humans need these rituals, and ways to connect in the midst of what can be seen as a world of separateness. This is about hope, agency, and interaction design. It's about surprise, delight, and deep belonging.

In 2021, I gave a talk that showed my thought process when approaching emotional connection using shared AR. The following descriptions highlight the potential, too, in wearable spatial computing devices that operate using a passthrough view system where you can see and augment your physical surroundings and environment.

It's a simple premise and the key to why this chapter on digital wellbeing and relationships is such an important bead on the necklace: For healthy relationships, which foster wellbeing, to be mediated by AR technology, something must be "shared." There needs to be an element of sameness in AR. I've delineated it into three main categories:

- **Same place, same time:** AR can allow us to have copresent experiences in live time together. You are physically in the same location, and you experience AR together. For example, a group of friends walking around a physical park playing Pokemon Go.
- **Same place, different time:** AR can enable objects, experiences, audio, or any data to exist in that same physical environment, much like geocaching. You can experience a connection in relationship to

someone else through discovering shared augmented objects or leaving messages for each other at different points of time.

- **Different places, same time:** AR allows two people to be in two different physical environments, and they can connect in shared time. For example, you could have an augmented reality coffee together, where you are both in your respective homes sitting at your own kitchen tables, and each sees the other's avatar or hologram sipping coffee opposite you, spatially placed in your own home even though they are physically elsewhere. The shared copresence connects you!

It's important to realize that the whole reason that AR relationship building can be so powerful is the **attention** factor. It allows people to connect and pay attention to each other in surprising, fun ways, sharing objects and experiences that bring them closer without the friction that can sometimes be associated with VR. Our sense of flow follows our focus.

It's critical to be intentional about how we use augmented reality to add layers, mapping, interaction, and additional information and context to our physical reality. The world is already noisy enough, so it should not be about adding more volume.

It's critical to be intentional about how we use augmented reality to add layers, mapping, interaction, and additional information and context to our physical reality.

The "ambient" part of these interactions can also transform into active engagement, as we wish. The beauty here, whether in spatial or augmented, is that the expectations and rules of engagement are entirely different than those in the physical world.

"The Call Is for More Play": Spatial Networking

Those of us who have been interacting in spatial environments for decades know this well. Andy Fidel is one leading voice in the industry who expertly describes social collaboration and sets the stage for interactions that are the essence of mindful media. Fidel coined the phrase "spatial networking" and says, "The call is for more play and quality human time. It's about having the space and time to ideate, debate, and innovate alongside fellow thinkers. There's a strong longing for deeper intimacy, transformative moments, and the sense of being part of a larger community. . . . Spatial computing and XR play an important role in bringing my vision to life. They encourage diverse perspectives, create safe spaces for exploration, and empower active participation, fostering growth and experimentation. They have the potential to enhance and expand our reality."

We're going to discuss this more in the chapter exclusively devoted to XR, spatial computing, digital games, and the power of play. This is a taste. I like to focus on relationships and intimacy that is not artificial, even as it's mediated across a bridge of technology. I say "spatial is special" and distinct from physical in interaction expectations. How I would define some of the differences and opportunities in spatial engagement include the following three basic features that might be different than physical interactions. These are purely described from my own experience.

Why Spatial Is Special

- You are encouraged to look in 360°, as well as up and down, while you have a conversation. There's not the same eye-to-eye gaze that is expected in physical interactions because those eyes are inhabited by an avatar. (Note: There are also spatial experiences where you are a glowing ball of light, or not even embodied by anything. These are also worth considering.) Some people are very avatar focused, and others are not. I find that in most worlds, avatars will evolve and shift, and the beauty is in a literal and philosophical nonattachment to form. You can be very connected with a person who is with you in a shared experience in spatial without needing that face gaze and eye contact that is a norm in a lot of cultures in the physical world.
- You are "free to explore" within your spatial location. It's not seen as antisocial to "wander off" as you explore a social space that is part of a social interaction with someone else. Of course, if your audio is also spatialized and set to 3D mode, that means you might have a humorous moment where the other person cannot hear you. If spatial audio is not enabled, this means you can hear your friend, and potentially everyone else, throughout the space/instance, regardless of location. It sounds simple mechanically, and the expectations here become so different because of this. Getting lost is entirely safe and in many cases expected as part of the journey.

- Silence is okay. It's okay to be quiet. Watching and observing and then gently participating is completely socially acceptable. Ah, deep breath of relief for all of us who have ever felt that "awkward cocktail party" feeling in the physical world, where we struggle to make small talk and feel the intimidation of stares if we are standing by ourselves at the edge of a room, a "wallflower." In spatial, ironically, we cannot even use the term wallflower because the wall could be the ceiling or the floor . . . or the outer edge of the galaxy! The edge becomes the heart too. Expectations are upended, and everything is possible, as long as mutual human dignity is honored. That's my tenet. In social spatial interactions, it's not seen as antisocial to take your time. First off, no one really knows if you are experiencing a technical difficulty, so we have to be extra accommodating and understanding anyway—we have all been there when tech has failed us. Besides that, it's an "opt in" environment where it becomes your freedom and your choice to use layers of voice, movement, objects, emojis, and the like to communicate. There are many ways to connect.

When people gather together across distances of space, the majority of us genuinely want to connect in new ways that are all about shared emotion and authenticity. I keep in mind that everyone I meet in avatar form is a whole, embodied person, with a range of emotions. Now, of course, there are NPCs (non-playable characters) that are programmed, and of course there are also AI integrations, and with all of that in mind, rather than emphasize divisions, let us use all our best knowledge to enhance the experience itself. We can celebrate a connection that is not only possible but probable and palpable given the opportunities in this medium!

Thinking even more deeply about ambient copresence and opportunities to treat social isolation, loneliness, and depression, research is underscoring many opportunities for digital wellbeing to play a role in fostering connected mindful media that has a direct impact on health. The copresence phenomenon is facilitated by digital technologies and has implications for various aspects of human life, including mental and physical health, social connection, productivity, creativity, imagination, belonging, and connection over loneliness.

Getting Granular with Digital Copresence That Forges Connection and Heals Loneliness

We've already explored many examples that flesh out why digital copresence, ambient and active, has significant effects on overall health and wellbeing. It's one of my discoveries through experience design that I prioritize. Here are some additional ways that it "plays out" specifically in shared experiences involving digital copresence:

- Social media platforms, both in 2D and 3D worlds: Users can extend their social interactivity with acquaintance networks in a superficial dimension, enabling a form of digital copresence. Users experience a peripheral yet intense awareness of others' presence through status updates, shared content, and the visibility of online/offline statuses.[4]
- Remote work tools: The idea of connecting remote offices or teams through continuous video links or shared digital spaces has been explored to create a sense of digital copresence. This concept involves using digital platforms to simulate a shared workspace, where one can have a low-resolution understanding of colleagues' activities, energy levels, and availability without direct spying or intrusion. This approach aims to replicate the social dynamics of a physical office, enhancing motivation and productivity by making remote work feel more connected.[5]
- Gaming environments: Certain video games have implemented features that allow players to experience the presence of others in a shared digital space. For example, games where players see "ghosts" of other players flickering in and out of existence, communicate through leaving messages, and interact in constrained ways, such as through movement and gestures. This gaming equivalent of copresence enables players to be aware of and interact with others while still having their own individual experiences.[6]
- Social media and microblogging sites: These social post platforms contribute to ambient awareness, a form of copresence, by allowing users to develop a sense of awareness of their online network through constant exposure to social information. Despite each bit of information seeming like random noise, the incessant reception of updates can

> amass to a coherent representation of social others. This peripheral awareness develops from fragmented information and does not necessarily require extensive one-to-one communication.[7]

These examples illustrate how copresence is facilitated through various digital platforms and technologies, enhancing the sense of connection and presence among users who are physically apart.

Researching these topics made me realize three things that I appreciate about *asynchronous* digital copresence, because, for me, I still experience that "thrill" when it is not in exact live synchronicity:

1. **Asynchronous digital copresence lets me feel as if I am receiving a present that transcends time**. I toggle back and forth in my mind, experiencing the moment that someone left or placed the gift for me, and simultaneously travel to the present moment of my own discovery, feeling it even more deeply. It's as if a golden thread has traveled from them to me, and we exist together, apart from constraints of time. I feel this every time I participate in geocaching. It's not two separate moments. They become one through that intentional serendipitous transmission. I also feel this way when I read great books, especially ones that are written by authors who are long deceased. Their words really live and travel off the page.
2. **Asynchronous ambient copresence lets me take my time.** I am not rushed, and expectations for my behavior or response dissolve. It's a meditation of appreciation. I can hold the moment. My favorite games often behave this way: they have intrinsic motivation and their own rewards and beauty, and there is no countdown clock to rush me, no linear path.
3. Similarly, capping off the triad, **asynchronous ambient copresence reminds me of both its ephemeral nature and its endurance by being there for me when I need it.** When I come across it is the exact moment I need it. This is why certain friends and I leave notes for each other on channels like Discord without clamoring for immediate attention. We are in separate time zones; we have events interspersed throughout days that are never the same. Each day is to be appreciated. The notes endure or dissolve.

> The larger connection persists. It's a lesson for me in staying open, in letting ideas wash over me like waves.

Thinking about the positives and negatives of social platforms, I know that I love to use them to share information and meaningful data and discoveries with others, and I start to resent them when they mimic popularity contests or "cheering" events. Does that make me a pessimist or cynic? Certainly not. That said, I know the feeling that creeps up immediately after I post something, or maybe 10 minutes later (ha), when I'm tempted to return to the post to check and see how many likes it has or how many comments have been left there. Then, I feel an expectation or urge to engage with those engagements, to volley back a thought. That's how the algorithm works, is not it? It draws us in. And worse, it's not seen as a meaningful interaction if it's not reposted to other people's feeds. Then people become experts at playing social media algorithms instead of focusing on the genuine content they shared in the first place. The art becomes artifice.

There's a long game here and a renaissance about to come when we speak up for ourselves and act in the name of digital wellbeing. Let us be aware of the game social media is playing at our expense. There is dignity to consider.

Digital Wellbeing, Relationships, and Gaming

Talking with Dr. Guy Winch, a prominent figure in the field of psychology and emotional health, illuminates how digital wellbeing can be approached in an integrative way, without viewing technology and some social digital activities, like gaming, as negative. Winch's work emphasizes the importance of emotional hygiene and health,[8] which involve addressing emotional pain with the same diligence we apply to physical injuries. The principles he advocates for have significant implications for digital wellbeing, specifically the impact of digital technologies on people's mental and emotional health. In the context of Winch's work, understanding and applying emotional first aid can be particularly relevant in navigating the challenges and possibilities posed by the digital world.

I asked Winch about an article he wrote about gaming and technology for *Psychology Today* in 2019. Some of what we shared together leads us fluidly into the next chapter, where we talk about gaming. The article[9] led me to think about rituals and coming back from a difficult time of separation during the pandemic.

When I asked how he approaches connection, and why was he first interested in writing about gaming, emotional health, and relationships, Winch shared,

> *"In general, with emotional health, it's about the idea of how we integrate our virtual lives with regular lives. I first started talking about this around 2010, and in my first book, The Squeaky Wheel, I even had some references to Second Life. I definitely sense that sort of virtual world gaming is where we are going. And I, being a fundamental optimist, always look for how something can be helpful and enriching, rather than purely unhealthy."*
>
> *"The minute ChatGPT came online, my first thought was optimistic, even in its current primitive form. When I started having a conversation, I began asking it about itself, 'Do you refer yourself in a third person?,' just trying to get the parameters around how it has conversations. There are so many very lonely people who do not have someone to talk to. . . . Even in primitive form, there's something that can be done there. I'm always looking to see what new technology can do to enhance our wellbeing emotionally and our quality of life, and I think there's so many options."*

I wanted to hear from Winch about emotional connection and gaming, asking him, "In what ways do you think games can build empathy? What cautions or advice would you have for game designers who advertise their games as social emotional?"

Winch, with a nuanced understanding of the subject, delineated the concepts of empathy and perspective taking. "Empathy is what it feels like to walk in another person's shoes. Well in virtual worlds, you can actually put them in the shoes and have them walk around, not just do a mental exercise, but literally have a visceral experience," he explained, emphasizing the potential of VR to provide a substantive experience of another's life.

I responded to Winch's observations, acknowledging the complexity inherent in these virtual experiences. I shared, "I'm careful as I work in the space to avoid generalizations and acknowledge every individual's complexity and multifaceted identity." The wisdom of Chimamanda Ngozi Adichie on the danger of a single story and the risks of forming oversimplified narratives from singular experiences applies strongly here.

Winch agreed and pointed out the scarcity of varied perspectives in gaming narratives and cautioned against the oversimplification of complex identities. He highlighted the importance of considering individual emotional intelligence in the design of such experiences. "We have to account for people's emotional intelligence because that is going to really be a significant filter by which they take away or do not take away; they understand or do not understand; they understand or misunderstand and misconstrue . . . ," he stated, suggesting the potential role of AI in adapting experiences to the emotional acuity of users.

The conversation took a personal turn as I shared my initial attraction to VR, not for its technological prowess but for its unexpected tranquility, reflecting creativity, and wonder. "One of the reasons I initially saw such potential with virtual reality earlier on is because it's such an intimate and powerful transformative experience," I shared, describing the serenity and expansiveness of a VR moon garden I designed and built to smooth out the day's rough edges. "It's a place to share stories, to involve poetry, and to experience creative freedom and gain a sense of peace."

Delving deeper into the subject of game design, I asked about Winch's perspective on the advertising of games as tools for social and emotional development. He emphasized the necessity of empirical evidence. "The scientific experimental method is where it's at," he declared, pushing for empirical evidence to support the purported benefits of gaming experiences.

I then asked Winch about the therapeutic potential of social gaming. He recounted transformative stories of connection and belonging facilitated by multiplayer games. "I've worked with people for whom that was transformative. It took a very isolated, lonely person and gave them a sense of belonging. . . . They connected through this game, and that's how they created a tribe around themselves, which is so important," Winch shared, affirming the deep social significance of gaming in creating communities.

Finally, our dialogue shifted to the burgeoning intersection of AI and emotional support. Winch envisions a future in which AI could act as empathetic companions, offering support and reflecting the basic therapeutic interventions that are accessible in the initial stages of therapy. He speculated on the capability of AI to provide emotional regulation prompts, integrating seamlessly into daily life as both assistants and confidants.

As our exchange concluded, I sought Winch's definition of digital wellbeing, to which he responded with a question that reflected the complexity of the concept, asking whether it pertained to the impact of the digital world on one's wellbeing or the experience of wellbeing within a digital context. I then defined it as the integration of digital devices into one's communication, lifestyle, habits and relationships, mediated by technology, highlighting its contextual and psychological dimensions. I emphasized wholeness and the possibility of heightened connection and creativity. "It involves applied imagination as well as design," I clarified.

Winch concurred, recognizing digital emotional wellbeing as a multifaceted and already present aspect of modern life, with implications for our self-perception both in the real and virtual worlds. Our discussion, comprehensive and probing, touched upon empathy through gaming, the subtleties of virtual experiences, the dynamics of digital identities, and the anticipatory integration of AI in daily life. Winch's thoughtful perspectives, coupled with our shared reflections, illuminated the multifaceted relationship between digital advancements and human emotions, underscoring the importance of digital wellbeing as an essential component of contemporary existence. Through this dialogue, a detailed picture emerged, highlighting the profound influence of digital interactions on our lives and the ethical, emotional, and cognitive dimensions that game designers and users alike must navigate.

This conversation with Guy Winch was revelatory in many ways. Context drives everything, and content is no longer king without the context that shapes it and the intention and attention that guide how we approach it.

This chapter's bead on the necklace is all about connection and how meaningful relationships can be supported by technology. We look at existing views openly, yet consistently strive to transform them through redirecting our attention toward wonder and awe, which are a priority in our design for work and personal life. This chapter gives you even more ideas about ways to employ this new lens and use it to reach higher levels of digitally integrated wellbeing.

In the next chapters, we will look specifically at video games, emotional design, VR, spatial computing, and a whole array of immersive digital media that can be mindful and resonant, in the name of wellbeing. It's all about the design, the intention, and the attention.

Personal Reflections

1. What are you most concerned about related to the future of relationships? What brings you a sense of hope or determination?
2. How do you choose to share meaningful time and attention with someone else? Can we do this using technology as a bridge, or is in-person always better, in your view?
3. Have you ever experienced "artificial intimacy" or being "alone together" where you were with another human but felt disconnected because of your attention split on digital devices? What did it feel like?
4. What do you think of digital copresence and some of the AR and spatial approaches discussed in this chapter?
5. What are your thoughts about using AI as an empathetic companion, mentioned in this chapter as an assistant, not replacing a human but supporting overall access to emotion support and therapy?

Reflections for Leaders

1. Are we living in "the most awkward time to be human" in your view, when it comes to 2D technology? What have you noticed about the ergonomics of the nature of digital work?
2. Through this chapter, what new strategies do you have for ways that technology can forge genuine connections and relational trust for you and your teams?
3. How could "the call for more play" and spatial interactions dynamically shift the nature of workplace design and strategic priorities about ways to interact and share ideas?
4. How have organizational strategic communication priorities, expectations, and distributed workforce models contributed to aspects of digital fatigue? This includes overwhelm from online meetings, disruptions, and other sources of digital friction. Using ideas surrounding ambient copresence, how might you shift workflows and ease these frictions?
5. How can all technology, including social gaming and creative applications, bring us in better connection with ourselves and those we work and collaborate with? The sky's the limit, and this is your time to reflect and dream big.

Extensions for Further Exploration

When we think about loneliness as an epidemic, and the opportunity for digital resources to form more connection points for us and those we care about, including our personal and work relationships, it seems as if there is a tremendous source of hope evident through this chapter. Wonder-rich digital wellbeing provides many ways for agency and new forms of connection. It's the AI component that can be disorienting and concerning, and the "handoff from AI to human" in therapy integrations needs to be explored more deeply. I think there could easily be confusion here and blurred lines when it comes to humans subtly and subconsciously anthropomorphizing the machine. There's more research to be done. Perel and Turkle make great cases for connection and attention, and technology can help us. Human attention is a part of relationship building, and if our attention is not nudged toward each other and the beauty of the natural world around us, where will it be placed, and with what effect? How can imagination and wonder play a supportive and animating role? These are among the extensions to uncover, and they will continue to illuminate many facets of digital wellbeing.

6 Thriving with Digital Wellbeing: Approaches to Learning

In this chapter, the bead on the necklace is learning, which involves mindset shifts. Humans are lifelong learners, and we are living longer than ever, thanks to technology advancements and greater focus on wellbeing. We cannot learn if we are in a state of threat or unease, understandably. And we learn best when we are creatively inspired, connected with ourselves and others, and flourishing in collaboration and compassion. Technology is essential to supporting these types of imaginative ecosystems.

In this chapter, we'll talk about some of the spaces and places that are infusing wellbeing practices directly into the models for learning. The digital design is a part of their approaches, which take different forms. Many of these organizations are conducting meta studies of their own and are uncovering incredible findings through their investigations. Some are education centers; others are research arms and agencies that are developing ideas and connecting all of us committed to the cause. We'll shine a light on some use cases of digital wellbeing in action here, identifying some of the terrain at this stage. We've already mapped a lot of our own methodology, including how I approach teaching Digital Wellbeing by Design, with a focus on building healthy relationships mediated by technology. This can serve as a frame and a base for these conversations and investigations of our own.

What It Means to Thrive, and How Stories Play a Role

When I spoke with Emily Weinstein, cofounder of the Center for Digital Thriving at the Harvard Graduate School of Education,[1] what came up again and again are the challenges and opportunities presented by technology in the lives of young people. Weinstein, alongside her colleague Carrie James, has embarked on a mission to ensure that individuals, particularly youth, can navigate the digital world in ways that promote their wellbeing and empowerment.[2]

Weinstein and I talked about research-informed practices and new ways for learning communities to honor different backgrounds and influences that play out in our approaches to tech. Her research, often conducted in collaboration with James, explores the complex dynamics between youth and social media. Their 2023 book, *Behind Their Screens: What Teens Are Facing (and Adults Are Missing),* dispels common myths about teens' digital engagement, offering insights into how adults can better understand and support young people in the digital age.[3]

One question I had for Weinstein stemmed from thinking of a "center" as an accessible place for everyone, at individual, group, and society levels. The naming gives it a nucleus feeling, as a hub dedicated "for" something. It is functional and purpose-driven. I believe that words have power, and as much as we should not dwell on semantics, we should give thoughtfulness to the names we choose. I asked Weinstein if the center addresses people first at an individual level or a larger group level? I think this could also be framed as an "inside-out or outside-in" question about intention and mindsets, as we seem to have individualized approaches to practices in mindfulness and wellbeing in Western society.

When it comes to wellbeing and resilience, how does Harvard's Center balance and approach individual and collective views? Weinstein said, "My disciplinary bias is in psychology and human development.

My instinct is to start with the individual. But I have now worked very closely with my research partner, Carrie James, for over a decade, and she is a sociologist. I feel like I can engage this both ways. My high level goes back to (Urie) Bronfenbrenner. I think about nested ecological systems and individuals and contexts. And it feels like those interactions between the individual and all the different settings really makes sense to me as a framework. Trying to tease apart the individual from the context feels futile in a lot of these practical applied conversations. My instinct is to start with the individual. That is not to detract from the importance of the community. Both matter."

Weinstein continued, "When we started working with Common Sense Media, they had this digital citizenship model that was 'self, close friends, and community.' And it was three concentric circles, similar to the Bronfenbrenner's idea. At Common Sense Media, they call them Rings of Responsibility. And the idea was that they were trying to have students really thinking within and across the Rings of Responsibility in their digital lives and digital citizenship."

Bronfenbrenner, whom Weinstein referenced along with the Rings of Responsibility, has an ecological model that helps to explain the complex systems that impact human development. Because life is constantly evolving, never static, and highly influenced by context of surroundings and relationships, this model is a highly useful framework that can help address the "dynamic interaction between environment, societal, biological, and psychological factors."[4]

Personal history shaped the model's development. Urie Bronfenbrenner was born in Russia in 1917 and immigrated to the United States at the age of 6. His father, a neuropathologist, worked with developmentally disabled people. Urie witnessed his father's concern for the impact of social contexts on individuals, and this influenced his later groundbreaking work in developmental psychology.[5]

Bronfenbrenner received a PhD in developmental psychology and his early research focused on the development of children within their peer groups, which marked the beginning of his lifelong interest in environmental influences on human development.[6] After serving in World War II, Bronfenbrenner joined Cornell University, implementing theory and research designs at the frontiers of developmental science, laying out the implications of developmental theory and research for policy and practice, and communicating findings to the public and decision-makers. His congressional testimony in 1964 and subsequent work helped establish the federal Head Start program.[7]

Bronfenbrenner's ecological framework, known as the Ecological Systems Theory, transformed the study of human development. It emphasizes the importance of multiple levels of influence, from immediate family and school environments to broader societal and cultural contexts. His theory is received worldwide as a comprehensive approach to understanding human development, influencing research, policy, and practice.[8]

I've been thinking lately about our various life histories and cultural shifts in this modern age of increasingly fluid travel and family structures, tech advancement, and approaches to health and longevity. Are there ways to apply Bronfenbrenner's model to a digital wellbeing human bioecological context? Can we be intentional about using it as a conversation piece? How can we build in even more consciousness about broader environmental impact and our nonhuman living creatures we live with on planet Earth? Could we incorporate nonliving entities, such as AI systems, in how we talk about our bioecological development? What model can serve us well as we approach the mid-twenty-first century, this critical period of exponential AI development? Let us build on previous chapters' relationship considerations, now adding in an expansion of Bronfenbrenner.

Digital Wellbeing Layers of Connection: Expanding Outward in Impact

Common Sense Media's Rings of Responsibility is a framework that encourages individuals to consider the impact of their actions, both in the physical and digital worlds, by using the metaphor of "rings" that represent different levels of impact: oneself, one's community, and the world. This framework emphasizes the ripple effect of actions, highlighting how individual behaviors can have far-reaching consequences in both worlds.[9]

Let's imagine a new complementary model that addresses wonder and wellbeing. We can form a digital wellbeing set of concentric circles, applying those layers of connection to our strategic outlooks, looking at the interplay and impact of personal and work lives as individuals, leaders, and members of communities. We can call them digital wellbeing layers (see Figure 6.1).

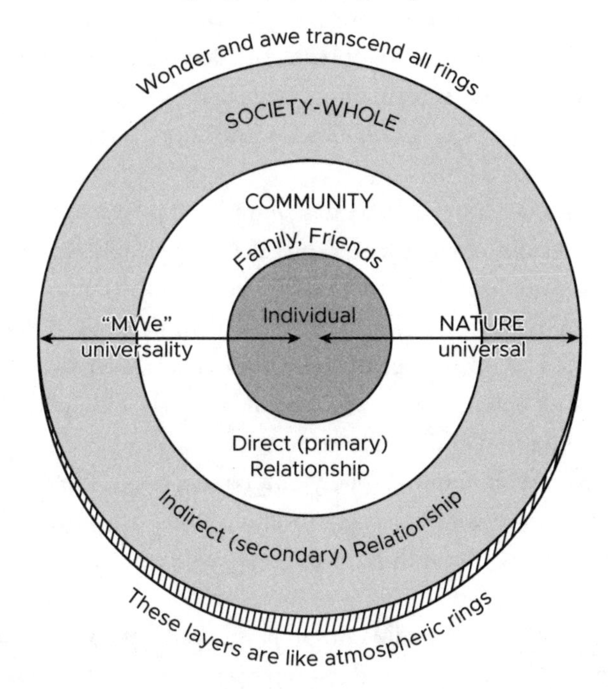

modeled after the "Rings of Responsibility" and the "Bronfenbrenner model"

©Caitlin Krause, 2024

Figure 6.1 Digital Wellbeing Layers as concentric spheres.

These layers can coexist simultaneously when we approach them using wonder and awe as guides. Creativity, curiosity, and imagination play an important role in helping reshape outlooks surrounding both tech relationships and broader life itself, in part through encouraging someone to take different perspectives, and to rethink assumptions about identity and habits in an open, curiosity-driven way. There is great freedom that gaming, 3D interfaces, wonder-rich environments, and music and somatic integrations can provide. It's all in the art of the design and the onboarding and post-immersive integration too. In studying the research about imagination and flow, I've found that reappraisal is highly powerful in these states of open wonder. An appreciation for the intrinsic value of experiences reminds people of the value of their own presence. In cultivating gratitude and self-compassion as part of our development of addressing habit formation and change when it comes to technology, staying flexible, adaptive, and courageous in our mindsets is invaluable for educators, parents, mentors, and learners alike.

Using this new approach to digital wellbeing, we can see the rings inviting us to dissolve attachment to one perspective and one fixed identity. While what is most proximate to us could feel the most important, there are wider circles of expansion that also represent a wholeness and that collective approach layered as extensions of individual. It's another way to approach community and belonging in the digital, natural, and imagination-rich worlds, with intention and a sense of wonder.

I asked Weinstein about empathy and connectedness in storytelling. In our conversation, we were referring to it as a shift "from eye roll to empathy" in social groups. Unwittingly, we can separate ourselves in terms of age groups, social groups, and all sorts of divisions, looking at each other with eye rolls instead of understanding. It's in part because we feel those divisions and have not created enough bridges. Separation can exist between teachers and students, parents and teachers, parents and students, and across peer groups—they can exist everywhere, and we can unwittingly reinforce our own biases if we are not intentional.

Weinstein addressed these divisions in detail and spoke about how to foster healing, empathy, and understanding. "This shift from eye roll to empathy is so important. How can we encourage teachers and mentors to do more of that? I think that you have this instinct as well, based on our conversation. I believe so deeply in the power of storytelling. I think that why we eye roll is because we have a lack of understanding. We're not leaning into perspective taking, asking what it looks and feels like for teens, and why this is hard for them. Perspective taking needs a developmental lens, considering what's actually important to you as a teenager. It's not just asking, 'What is the teenager thinking about, in this moment? What's the sort of explicit dilemma?' but also, 'What are these implicit developmental drives?' I do think that storytelling is such a powerful vehicle for helping us feel invited into someone else's perspective."

When Weinstein and James teamed up to write *Behind Their Screens*, they drew on perspectives from more than 3,500 teens to provide a powerful reframe of what teens are up against and what they need adults to understand. This book has a strong research lens, exploring the complexities that teens face in their digital lives, having grown up with social media. Both authors warn against a single-minded focus on "screen time" and emphasize the need for adults to understand the nuanced nature of teens' online lives.[10]

Talking with Weinstein, I was curious to hear her views about how all generations can access each other's stories and find relatability. I asked how groups can "come together and see past perceived differences that might create implicit bias?" Without hesitation, Weinstein said, "I think storytelling . . . asking powerful open-ended questions that create opportunities for us to share our struggles and experiences. In particular, those (stories) that might not be apparent or visible to people who are at different stages of life, feels like a really powerful bridge."

Our perspectives reflect essential truths about the power of connection and story. Throughout my global teaching and learning experiences, I learned about the power of framing meaningful questions. Looking at this conversation high level, it describes what everyone could be considering in their personal and professional communities: how to have better relationships with each other, how to think of digital media as a vehicle for connection and not the end product for consumption, and how we can creatively and imaginatively reframe the stories we might have unwittingly been holding about ourselves and each other. These are tenets we are upholding in this book, in our approach to "digital wellbeing." To be well means to thrive.

Being better listeners, staying open, and maintaining proactive hope, these are skills that need practice. The Center for Digital Thriving at Harvard is contributing to the field of digital wellbeing in a significant way through ongoing research, advocacy, and the development of practical resources. They are at the forefront of efforts to ensure that technology serves as a tool for empowerment and positive growth, particularly for young people navigating the complexities of the digital world.

As a side note, I especially value that neither Weinstein nor I seem to become too attached to words, although we are intentional about that naming process. What we *mean* is more important than any buzzwords, which might be loaded with connotations and expectations. Perhaps it is time to shake up language to form something new when it comes to wellbeing.

My prediction and hope is that younger people will help us develop that new language and wellbeing literacies along with new tools that become more about what we are choosing than what we are rejecting. It's necessary to release the past in order to embrace what will come. As André Gide said, "You can never cross the ocean until you have the courage to lose sight of the shore."

A Necessary Shift in Mindset

This new time takes great courage. There are many models for how school can operate differently, with mindfulness, social connection, and thriving as priorities. My book *Mindful by Design* (2019) took its inspiration from my decade-long experience as a full-time teacher in middle school and high school classrooms, working to directly infuse social-emotional and mindful imagination-rich practices into learning environments. When I founded the consultancy MindWise, leaders and educators were telling me they "did not have time" for creativity, wellbeing, and collaboration, in or out of their work practices, and they did not know how to apply it in context and content, so I created the book to offer both theory and practice.

A few years later, I wrote *Designing Wonder* because my work in virtual reality and spatial computing experience design underscored the importance of prioritizing design strategies that use wonder and awe in forming the foundation of our mindful attention and its quality of approaching technology. Now, focusing on wonder-rich digital wellbeing prioritizes thriving with imagination and wonder in a tech-saturated world, showing paths to new levels of transformation instead of shunning what will ultimately become further integrated into our lives. We can learn how to have more full human lives through these methods and strategies, in collaboration fueled by curiosity and openness.

This chapter maps some of those discoveries on the journey and is certainly the beginning of further expansions on the topic of lifelong (long life!) learning and thriving, which affect our personal lives, work lives, full lives. The challenges to come also hinge on our ability to discern quality and authenticity when it comes to technology. At a recent Center for Humane Technology (CHT) event, "How Can We Tell If a Technology Is Good for Us?" that Randima Fernando hosted, a group of us imagined a mindset shift.

> *"The most important piece is the mindset or the paradigm, from which everything else is derived."*

Fernando cautioned, "When we deploy technology, it lands in a complex socioeconomic system that's constantly evolving. Our system has various pieces that control how it functions. So it's got goals, it's got decision-makers' rules, information and material flows, positive and negative feedback loops, incentives, and taxes, and so on; there's a lot of complexity. But the most important piece is the mindset or the paradigm, from which everything else is derived."

Education is one of several societal systems embedded and sometimes entrenched in challenges related to mindset. And it's a media issue, too. "Part of the challenge is that technology exports the mindset of its creators through its products and interfaces." Are the mindsets ones that embrace and honor thriving? Do we have an individualist or collective approach, or both? Embracing the layers of digital wellbeing, it's possible to see from multiple angles and have more plasticity of mindset.

Mindset concerns apply to product creators, business managers, education leaders, product designers, and beyond too, as learning as a system and even creativity can be viewed as a product created for consumption by those who experience it—the learners, the teachers, and subsequently every part of a community.

In the CHT discussion, we addressed what it means "to thrive," complementing conversations with Weinstein. "Does the product help users thrive? Our definition of thriving is our sense of what a good life is and what it means to be happy. From there, we derive our guideposts for our life. Think of them as principles that shape our behavior, including what we think, what we say, and how we act. So our definition of thriving is ultimately responsible for everything we do.... Every action you take is because in some way, you think it contributes to your thriving."

As I've shared, wellbeing and thriving seem inextricably intertwined in approaches and definitions. There's a certain meta level to the discussion, building on our previous chapters in these beads on a necklace: How often do we discuss sheer joy when it comes to wellbeing? Where does the idea of "play" with purpose fit into the equation? How about projects, passion, and peers? That said, there is space here for addressing comfort crushing and safe rigor, which I place among the top tenets of healthy growth and learning. We will talk about this more in-depth in the next chapter, when we discuss game design and the origins of "The Four Ps" and Seymour Papert's influence. While many people's life goals incorporate "happiness" first, I resonate with Papert and Fernando: "Our image of thriving also needs to have a place for challenge, effort, and pain. Thriving grows as we balance our relationship between pleasure, pain, and caring kindness."

When it comes to social media, dosage matters, as we can get caught up in dopamine spikes and hedonistic treadmills. "This is perhaps the most common place technology companies fail is they deliver a reasonably helpful experience at far too high a dosage, usually driven by business incentives," Fernando said. "Too much of even a good thing is harmful, so you have to make sure users do not overdose. So for example, teens spend an average of 4.8 hours on social media per day: 1.9 hours on YouTube, 1.5 on TikTok, 0.9 on Instagram. At these kinds of dosages, addiction comes into play. Our neural receptors get overwhelmed, and they get desensitized. Which means that we then need increasing doses of the same stimulus to get the same feeling. And so of course, users end up seeking satisfaction in ways that makes them less satisfied. They end up frustrated and confused about why they feel so unfulfilled."

It's actually this addiction loop that drives people away from resilience and balance, away from thriving. One way to ease this is by using "wise friction." This type of speedbump can break the addictive social media loop. It can be a simple pause, a button press or a prompt that can move a user out of an unconscious repetitive behavior. I see this wise friction come into play with the invitation of wonder-filled experiences across all forms of media, including immersive technology, giving a breath of fresh air when needed, and offering a way to break that addiction cycle. In this way, wise friction can be invitational and beautiful to experience instead of always focusing on boundaries and restrictions. This has motivated me to create experiences through my organization MindWise that serve as that imagination-rich pause, that wonder-filled wellbeing speed bump that allows people to refresh and recenter.

These methods and experiences I have built remind us of the possibility of creativity and awe in the present moment, the brightness all around us and within us. Right now can be a bright now.

Right now can be a bright now.

Reinventing Learning Worldwide

While we work to address many issues that impact learning and thriving with digital wellbeing, a focus on wonder and social-emotional connection are paramount.

I spoke about new models of learning with Iryna Nikolayeva, senior product manager at Learning Planet Institute (LPI), a Paris-based organization dedicated to reinventing learning for all ages (lifelong learning) through collective approaches. It was founded by François Taddei with the aim of focusing on interdisciplinarity, diversity, and innovation, relying on synergies between research and educational programs, international alliance, transformation of organizations, and digital ecosystems.[11]

I asked Nikolayeva about how they address the social-emotional aspects of digital learning. She said, "One of the things that's our credo is that we teach people to take care of themselves, of others, and of the world. So we have this fractal view." The staff and students use the same digital tools in project-based learning, and they often eat in the same spaces, encouraging social sharing, and bonds that go beyond pure work-related tasks.

How does Nikolayeva feel about technology and education working together for increased connection? "[Technology] can be a super powerful tool to connect people. I think physical interactions will always play a major role, and we should think about, for example, how to use technology so that the interactions are as interesting as possible. Matching people, for instance, who have similar interests, but who have complementary skills."

These ways of sharing and connecting are valuable, and it's important to break down barriers between learners and mentors, allowing for exchanges of stories and "low stakes" ways to express vulnerability and allow the process of learning to become more visible, rather than emphasizing the product. In environments I design for learning, whether for my university courses or for corporate teams, I make sure to encourage visible learning models and exchanges of formative ideas. I also give people chances to develop new models of framing questions, using structures to scaffold conversations. Digital environments and experiences can help facilitate these goals. The more that wonder and awe play a role in the process, the better the level of imaginative agency and expression, in my experience. Imagine how this can impact future learning and development when it comes to innovation, health and wellbeing, leadership and team trainings, and strategy sessions, as we work to overcome our organizational challenges, solve pressing global needs, improve communication and connection, and more!

There are many inspiring examples of learning organizations around the globe working to transform how we approach the next phase of humanity integrated with technology. We are in a phase of learning how to transform learning, and it's uplifted by wonder and connection. LPI has joined with UNESCO to co-build a learning society of public and private organizations capable of addressing complex challenges, working together to find sustainable solutions and ways to reach the Sustainable Development Goals (SDGs) by 2030. Related to digital wellbeing and wonder, I have been part of building digital platforms and immersive imaginative ecosystems enabling human-centered participants to rethink and encourage mentoring, peer learning, curiosity-driven growth, and digital wellbeing by design, allowing them to share their skills and initiatives to the benefit of the whole.

Summing Up the Future of Learning and Digital Wellbeing

The ideas we bring up in this chapter represent a global effort to address the challenges and opportunities presented by digital technology in educational contexts. I want to break them out of the silo of "academia" and encourage everyone to think of learning as a lifelong pursuit, one that can be enjoyed for its intrinsic value as much as for the opportunities to connect with the world, raise awareness, and increase thriving for everyone.

Next steps are asking us to consider lab approaches, learning sandbox-style, and sharing examples of what's been working—and what strategies need to be reevaluated and shifted—when it comes to connection and thriving. How can the Digital Wellbeing Layers of Connection inspire and inform us? What came before that is useful pedagogy to remember? Metaverses and digital playscapes are great spaces in which to explore, express wonder, and navigate complexity, in part because they are so versatile and manipulatable, simple to scale and customize, easy to dismantle and rebuild. It's time to take all of the frameworks and guides and *put them into practice*. I would suggest that every organization start by looking to the metaverse creators and moderators for advice, because this is the time to build those bridges, and XR leadership has ample expertise at both the technical arenas and the social emotional connection. It's time for extended learning and a new form of animated reality that hinges on this type of expression and connection. This is what we are about to explore more deeply in the following chapters.

Personal Reflections

1. Describe a learning experience that was meaningful to you, in terms of its setting, the teacher or mentor, and the overall way the learning took place. Reflect for five minutes, writing down everything you can recall about how it felt to be a learner in that example.
2. How did that meaningful learning experience involve technology? If it did not, how could it, mindfully?
3. What is your response to learning more about Bronfenbrenner's history and context-driven development models?
4. Growing up, who or what had the greatest impact on you when it came to finding your own path? Did you ever encounter "Rings of Responsibility" or concentric models that anchored your thinking about the impact of your own learning and actions?
5. What is your response to the statement, "Thriving grows as we balance our relationship between pleasure, pain, and caring kindness." Should our thriving goals be to increase comfort and reduce suffering? In what ways can adversity also build resilience?

Reflections for Leaders

1. Looking back at your answer to the first question in the Personal Reflections, do you think the way that you described that learning experience has anything in common with your current work experiences, either through mentoring you receive or mentoring you give? Why or why not?
2. How could the current lifelong learning system be improved, in your view? Is it designed with digital wellbeing as a priority? If not, what needs to change?
3. How are your current physical and digital work environments designed? Are there places and spaces for sharing non-work-related activities (such as eating together in person) that build communication and relational trust?

4. In what ways does contemplating the "Digital Wellbeing Layers of Connection" modeled after rings of responsibility influence your thinking about your individuality, your current work's impact, and your strategic goals?
5. If every experience is also a product, as we talked about with CHT, what products is your organization creating that encourage humans to thrive? What would your working definition of "thriving" be, after encountering the messages in this chapter?

Extensions for Future Exploration

Looking to build on this conversation, I would like to evaluate and highlight digital wellbeing programs in schools and public and private organizations, with student interviews and examples of their projects and learning focuses. This would be an exciting and enlightening topical extension. I would also like to collect longitudinal data on programming underway, looking to hone in on wonder- and awe-centered learning strategies that inspire imagination. This is also part of why I continue my work in this field: to bring curiosity-driven wonder and awe to experiences grounded in wellbeing, collaboration, and imagination.

7 Wellbeing Through Spatial Computing, Gaming, and Extended Reality

This chapter's bead on a necklace is all about wellbeing and mindful media. We will discover why immersive media is so powerful at increasing our sense of imaginative possibility and practical agency. Digital immersive media reframes our wellbeing. Let us examine how and focus on strategies we can each use in our daily lives. If you have never self-identified as a "gamer" or someone familiar with terms like "spatial computing" and "virtual reality" (VR), this chapter is also a powerful primer that allows you to learn more about immersive technology. By adopting a curiosity-driven, designer's mindset, we can learn and grow together, and evaluate how these forms of media are transforming the future and bringing about greater freedom and wellbeing worldwide.

Why Spatial Is Special

Interacting with spatial technology is undeniably a remarkable and often life-changing experience. You are part of a digital environment that transports you, that moves your body and mind physically and emotionally. This type of immersive experience can involve you representing a character, your avatar, in a 3D world, not a flat 2D one. You become part of the experience.

As I brought up earlier, experiences are how we encode emotions as humans, and the experience of emotion is what moves us. As Celia Hodent, emotion and game design expert, points out: "Certain design elements of tech facilitate certain emotions, and our emotions facilitate our usage of certain tech." We will hear more from Hodent in this chapter as we talk about the intentional immersive design shaping our behavior, which we can modify with wellbeing as priority. As Hodent states, "Conditioning is where experiences shape reflexes."

Mindful media is all about capturing that emotion in a package (a delivery method) of experience, which is usually delivered as a game with a story. Why is this? Stories have moved us since the beginning of time. We remember stories. We understand them without needing to be taught how they work. They are a way, as old as civilization itself, of transmitting cultural norms, belief systems, values, behavior guides, sensations, emotions, and new understandings. Good stories entertain us; great stories move us, inviting us to play the hero on the journey, coming to our own new understandings at the end of the experience, which is often the beginning of the next one. They not only move us emotionally, they move us viscerally, behaviorally, and reflectively, embedding their design sensibility into us, which will translate into our formation of principles and guide future actions. They move us.

Good stories entertain us; great stories move us, inviting us to play the hero on the journey, coming to our own new understandings at the end of the experience.

Ancient Immersive Tales

Imagine yourself as part of the civilization of Ancient Ur around 2000 BCE. You were not reading the story of Gilgamesh—you were hearing it out loud, likely surrounded by community. It was memorized in lines of poetry and delivered aloud in an engaging telling of the tale, likely combined with artistic props and musical accompaniment. Similarly, the classic story of Beowulf was delivered by a bard traveling town to town, as part of the oral tradition. Considered the first major story in English literature, with its story set in sixth century CE pagan Scandinavia, most scholars think Beowulf dates back to the seventh or eighth century CE.

Language has power and intentionality in reaching an audience in a specific way. You would have heard Beowulf in its ancient form, as a member of that ancient audience, and heard an echo in the language hearkening the past that you had a context to understand. It would have carried a significance and sense of belonging for you. It also would have taught you about your culture in a very particular way, imbued with the emotions of the sound of the language itself, its cadence and musicality. As an adolescent, hearing the story while surrounded by your peers and your elders would have conveyed so much meaning and intention. The next day, you likely would have reenacted parts of the story, creating a game with friends, engaging in a storied ritual. This is an example of a mindful media experience in the past, where audience became part of the immersive story as part of their cultural tradition.

Fast Forward to Modern Times

Similarly, our modern times that we experience now will be ancient to someone in the far future. How will they sense what our present moment was like for us and how we experienced media? What purpose did it serve us? Hopefully, they will learn that in our generations, in this epoch, we uncovered that the short-term profitability schemes of big companies did not lead to our best quality of life as a society. We were feeling culturally lonely, bereft, and depleted. We changed the narrative. Perhaps the story of our age will be that we upended what had been cultural media captivity and turned it into something mindful, empowering, and transcendent. Right at the perfect timing, as AI advancements were allowing us to live out our best quality lives, our new way of being favored openness and freedom, and this allowed us to chart a course for the future—their present—with imagination and insight. We then focused our attention on care for the environment, care for each other, and exploration of wonder and imagination, as we deepened our consciousness.

This is all a real possibility, and the choice is ours, right now.

How to Experience a Spatial Extended Reality Hero's Journey: The SCUBA Framework

If we use the Hero's Journey as a model to represent what each individual feels and encounters during their extended reality (XR) journey, it anchors our design considerations about the experience. I developed the SCUBA Framework in 2020 (see Figure 7.1), serving as a guide. Individuals are given mentoring and support, mentored through a journey where they safely cross thresholds into unknown immersive worlds that allow them to grow, connect, respond, and discover new meaning. As they transition back to the "ordinary world," one of the goals is that they have an applied transfer of knowledge and have formed new understandings along the way. This is an emotional journey with many opportunities for wellbeing to connect to human imagination, presence, and purpose.

XR invites us to feel a sense of wonder that is transformative. Imagine that you are joining me, about to step across the threshold into an immersive environment focused on developing your wellbeing. Here are the five parts of the "SCUBA" Framework mapped in this Hero's Journey cycle:

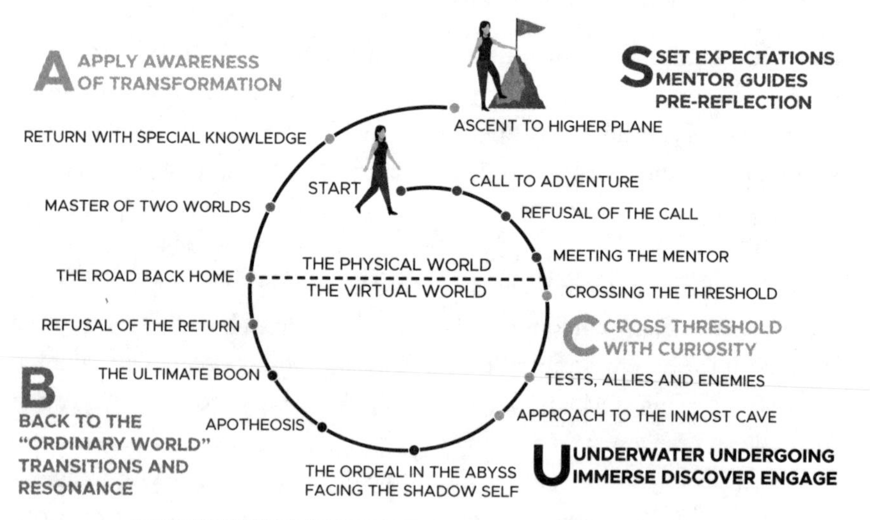

Figure 7.1 The XR Hero's Journey SCUBA framework.

"S" for Setting Intentions

You approach the experience just as you would begin a quest, with a feeling of heightened anticipation. You have received an invitation to encounter this, and it feels powerful and personal. I, as the guide, prompt you to set intentions about the experience and to frame its significance. I give you enough priming so that you feel empowered to cross the threshold, which is what is about to come.

"C" for Crossing the Threshold with Curiosity

As the Hero, you enter the new world with less inhibition and more curiosity. As the Mentor guiding you, I am responsible for your safety, and this is my prime consideration, along with evoking a sense of wonder for you. You, as the Hero, are invited to become more malleable, plastic, curious and eager. You cross the threshold, and you are surrounded by . . . what? What exactly? Imagine what could be around you. A snowy mountain range? An expanse of stars? An office meeting room? A wide summer field? A science lab? A battleground? A love story? The environment can be anything, and it will inspire your emotions and feelings in that space.

"U" Is for Under the Experience: You're Immersed in It!

You, as the Hero, encounter three key questions when you enter XR:

Where am I?
Why am I here?
Who am I?

I support this pivotal moment of entering the immersive space. As the Mentor, I am conscious of many ways to interact with you in the environment, either through an avatar that guides you or through the story I design that offers clarity, agency, and discovery. There's great intention and care involved on centering the experience around you and your transformation. When Heroes ask, "Why am I here?" they have clearly felt the call to adventure, and they need a strong sense of their purpose in XR. They need me to establish safety and trust, to clarify goals, and to empower them with agency in the experience. All feelings and needs are not just permitted but *invited*—it's all honored, valid, and connected in this authentic wellbeing immersive experience.

"B" Is for Back Across the Threshold

Before you know it, because you have lost track of time—another attribute of being in "flow state" and powerfully associated with immersive experiences—it is time to cross the threshold back to physical reality. My job as a mentor and guide is to help you do so with expertise and care. Have you ever watched someone wake up from dreaming? They blink their eyes; they slowly adjust to the senses of this world. Their mind is somewhere else; their consciousness has not yet arrived. I safely lead you out of the experience, gently guiding you back to the external physical world and consciousness. Otherwise, it could be a jarring return, and we risk losing the precious moments of deep reflection and integration. I have been supporting you, even if you are so absorbed in the experience that you do not realize this consciously. In XR, we need to care as much for the "coming out" of the experience as we care for the "going in" entry. They are complements, and they allow for the power of the experience itself to be truly transformative.

"A" Is for Applied Awareness of Transformation

The other reason to care for the "coming out" and the leaving from the experience is because, after this "exit," immediately beyond, lies the glowing moments in which you can introduce deeper reflection, encoding of meaning, and personal and group transformations. If the experience is profound enough, it is the *time after that experience* that will make a tremendous difference. Have you ever experienced something impactful—a book or a film, for example—that made you think and feel something deeply, and you wanted to share about it immediately after the credits rolled, or just after reading the final page? You were bursting with thoughts and questions, responses, and feelings, and you needed to share about it? It could have been that you craved a social outlet to share about it, whether over a meal with a friend who was sitting beside you in the theater, a discussion with a therapist, or a debate with a book club. It could also be that you longed for some silent, solitary time to reflect and let your mind play with all of the ideas you were exposed to in the immersive experience.

Positive Mentoring with XR

All types of reflection—shared and solitary—are encouraged and scaffolded in the full SCUBA methodology of immersive imagination. Following this framework is a way to increase freedom within the experience itself.

You are designing the frame and allowing for a deep dive into a realm where it's possible to reach new levels of understanding and awareness *if it's approached* with great care and attention to the integrity of the entry and exit.

When I design and guide an experience using the SCUBA methodology, it has much better resonance and meaning for the users. The group forms a community, and everyone feels safety and trust. Also, participants have a greater ability to find value in what they experience in XR and make that clear transfer back to their "ordinary worlds." It aligns with their sense of purpose.

I once had a powerful experience receiving XR mentoring at a VR film screening. I was using a VR headset to experience a documentary 360° film. If you have been to a show in a "dome" (think of Sphere in Las Vegas), this is like that, only the immersion is all happening in-headset. I went into the screening not knowing much about the film. When I entered the room, the guide who welcomed me was part of the film crew. She told me the film's length and a bit about the background. She made sure I was sitting comfortably, helped me to adjust the headset, and made sure the volume and focus were set appropriately. A Mentor needs to be there to guide these "first steps" to make sure the Hero is comfortable. From the Mentor, I knew that my purpose was to surrender and connect.

When I watched the film, I felt the emotions of what surrounded me in 360. Everything felt very real. After, when I removed the headset, my Mentor talked with me and asked some reflective questions about what I felt during the experience. I had time to process my thoughts. This scenario might not be feasible for other contexts. It gives an idea of what is possible when experiences are guided with care.

In the example I gave of being at the 360° immersive film screening, I wasn't embarking on a complex quest that required strategy or special training, yet I still needed to be guided across the threshold in a mindful way. If I were playing a more complicated game, or entering a VR social world where there would be different levels of interactions possible, the role of the Mentor could become even more involved, too.

Guiding XR Remotely

Because the VR experience itself is "in-world" (meaning, in the special immersive digital world), the Mentor might not be physically in the same location as the Hero before entering the VR realm.

In other words, if participants engage in immersive spatial experiences from their home location, I, as a mentor/host/guide will not be with them physically to adjust their headset or teach them about controls and setup. Still, I have found ways to mindfully and comfortably "onboard" them to the experience, offering safety, guidance, confidence, and compassion as they have a fully transformative experience. There are many ways to establish trust and create the sense of safety and assurance that users, aka the Heroes, need during their Journey.

The two "thresholds" of entry and exit involve putting on the headset and taking it off, yes, and they also involve a *lot* of safety and trust that might be underestimated at first, until you try VR yourself. Imagine how challenging it can be to close our eyes without full knowledge of what's going to happen in the next few moments, as our vision of the external physical world is replaced by a new digital world. Many of us might not be comfortable closing our eyes, for many different reasons. We need to honor our own personal safety and have choices when it comes to how we cross thresholds.

With the continuum of XR, especially VR, there is exponentially greater risk with this "close your eyes" invitation, because in VR you are essentially being asked to "close your eyes" to the world as you know it, putting on a headset that will replace all of your contact with the physical "real" world, replacing that world with sounds, images, and movement. You are entrusting your body and mind to someone else's care. That act needs to be considered and respected. Notice that even if the immersive XR experience has a pass-through feature, it's still a different experience to have a device on the head, obstructing the eyes. Great care is still needed, and individuals are encouraged to check in with themselves along the way, responding to what is best and most needed for them to have a positive experience.

Support through this process, and an invitation to use wellbeing and wonder, is part of what can make immersive experiences transformative, whether they are company trainings, meditations, explorations, games, or something that blends all of the above.

As part of XR mentoring and trainings, and our examination of the immersive storytelling practices that move us and engage us, let us look a bit more deeply now at some game design strategies and distinctions between games and immersive media and experiences. This chapter is that bead on the necklace of wellbeing that truly dives deep into **meaningful engagement**, and many would argue that there's no better way to engage the full range of senses, body, and mind, with all of our emotional sensibility, than games. Let us investigate why.

Game Design, and the Four P's Seymour Papert Mapped

I am a longtime fan of Seymour Papert's beliefs and teachings, among them *Mindstorms*. He encouraged me to "think more about learning and learn more about thinking" in my own design and teaching practices. I work to empower students in myriad ways, and my motivation evolved to become embodied in those Four Culture Cornerstones of Dignity, Freedom, Invention, and Agency that we discussed in Chapter 2. Everyone comes with their own set of stories, and *we contain multitudes*, as Whitman eloquently said.

Constructivism—the education idea that children are not passive vessels of knowledge, but active cocreators, integrating new findings with their own life experiences, contexts, and understandings—was a classic belief of Piaget, Vygotsky, and Steiner alike. Papert was also a proponent and built upon these learning ideas with his "constructionism." Game design can take a more constructionist approach in the way of inviting a player to build and create, thereby tinkering and interacting with the experience in a personal and physical way that is rooted in their own personal identity and history of meaning and sensemaking.

Papert was the source of the Logo programming language, inspiration for LEGO Mindstorms, and part of the vision behind the creation of the MIT Media Lab. He mapped out the "Four P's" of "projects, passion, peers and play" that are the basis for learning coming alive.

He was a pioneer in the fields of AI, machine learning, and educational technology, with a particular focus on how children learn and interact with technology. His beliefs about digital wellbeing and game design were deeply intertwined with his broader educational philosophy, which emphasized that importance of constructionist learning—a theory that suggests people learn best when they are actively involved in making tangible objects in the real world. I'm wondering if those "tangible" objects could be ones we perceive as tangible, too, when we create them in spatialized 3D worlds in XR.

The Power of Play

Seymour Papert passed down that legacy of constructivism that includes the value of "hard play," engaging in activities that are challenging and require effort, yet are enjoyable and intrinsically motivating. This concept aligns with a "flow state" with challenge/skills balance and an active, enjoyable, and meaningful process.

Papert believed in asking students open-ended questions and giving them freedom to experiment, explore, and come up with both their own solutions and their own further questions. He wanted them to question and understand the "how" of their own learning, which mattered even more than content knowledge. He believed in the power of objects to tell stories. I use that belief in practice in my experience design.

We all love to experience the world this way, right? The beauty is that we are never done learning and growing, and this lifelong activity of hard play can stay with us, affecting how we expand our ways of thinking about ourselves and the world. The human brain has plasticity after all, and science is now showing us that the adaptability and resilience we can develop are open to us at every age and stage. This fills me with a sense of hope and possibility.

These ideas directly impact the way that I approach game experience design and the ways that we all can open up to ways of thinking that are not so new, as they are reminders of truths that are inherent to our core identity and wishes as humans: to come alive with wonder in our interactions with media.

We need to *feel* in order to care. And we care deeply when we are involved in a game that is emotionally sensitive, empowering, and authentically aligned with our intrinsic motivation and belief systems. It can uplift us and empower us to form new understandings about ourselves and the world around us. On top of all of this, how often do we mention "love" as a key ingredient in meaningful experience design? Love and connection are present as part of the core of the experience, when it is designed mindfully.

We need to feel in order to care. And we care deeply when we are involved in a game that is emotionally sensitive, empowering, and authentically aligned with our intrinsic motivation and belief systems.

This type of design is exuberant, joyful, emotion-rich, and passion driven, not a linear "subject → goal" sort of path. Papert is not alone in these belief systems and practices; he is among a whole set of people who believed and dedicated themselves to transforming the experiences of game design and play. What is remarkable, and appropriate, is that Papert was also a pioneer in integrated AI applications that model and teach us how we learn best.

Intelligent AI-informed tutoring systems and human interactions with digital immersive forms can be mindfully approached using some of Papert's work and beliefs as a standard. We will build on them here.

On Digital Wellbeing and Game Design

Papert did not write extensively on "digital wellbeing" as it is understood today, but his work implies a belief in the positive potential of technology to enhance learning and creativity. He contended that problems with educational computer usage arise when machines are isolated from the learning process rather than integrated into it, as they should be for lifelong learning.

When children play games like *Sim City* or *The Oregon Trail,* Papert believed they learn about subjects such as urban planning or the American West as a by-product of the play. He saw the potential for games to teach creativity and innovation, by design, enhancing abilities in the players that they can transfer to "life skills" useful in jobs and work. Just think about that concept applied today, when the World Economic Forum has cited "creativity and imagination" as the top skills needed for the future of work, especially in this age of AI. How can we use these understandings to increase our prioritization of fostering imagination and a sense of wonder and awe? The answer, in part, can be in how we prioritize and approach game design and immersive play, for adults and children alike.

The rise of AI and machine learning is disrupting jobs in a way that urges us to look for new ways of approaching training and future of work. The traditional job trainings of the past will not work for the future, and fear or stasis are not the answers. Papert's legacy is impacting our lives right now. In our current age of AI development, we must increase our investment in ourselves. Investing in ourselves looks like developing our imaginations, and we do this through wonder. We do this through games and play that teach us more about creative problem-solving, that empower us to develop computational thinking skills, that train us to respond to what is emergent, that facilitate adaptation and wondering and scenario-building of "what if." This is where curiosity thrives! Play also allows us to explore and develop who we are, teaching us how to be part of collaborative work with others.

When this type of wonder, imagination, and play are built into our systems of work, integrated with AI, this is how we change the future and empower human-centered ways of being in powerful AI-driven landscapes.

Metaphorical Links and Overlaps

Papert's work in AI and machine learning and his ideas about digital wellbeing and game design are metaphorically linked through his **constructionist approach to learning.** He believed that technology should not be used to program children but rather that children could use technology to build and express their understanding of the world.[1] This philosophy is evident in his development of the Logo programming language, which allowed children to control a "turtle" to create graphics and learn programming concepts through exploration, unstructured play, and tinkering.[2] These are lifelong skills.

Technology should not be used to program children, but rather children could use technology to build and express their understanding of the world.

Seymour Papert was a pioneer in the fields of AI, machine learning, and educational technology, with a particular focus on how children learn and interact with technology. His beliefs about digital wellbeing and game design were deeply intertwined with his broader educational philosophy, which emphasized the importance of constructionist learning—a theory that suggests people learn best when they are actively involved in making tangible objects in the real world.

The development of intelligent tutoring systems took their inspiration from here, with insights into children's thinking and learning leading developers to recognize that computers could be used not just to deliver information and instruction, but also to empower humans to experiment, explore, and express themselves. Papert's work in AI and machine learning laid the groundwork for the conceptualization and development of AI today and **intelligent learning design, emphasizing the importance of active, experiential learning in the educational context**.

Papert's contributions to AI and machine learning, and his ideas about digital wellbeing and game design, are rooted in the constructionist theory of learning. I'm reemphasizing this here, layered on top of our previous chapter about learning, because it has so many implications for our digital immersive experiences that can and will shape the future of wonder-infused well-being for humans.

Technology, when used as a tool for active construction rather than passive consumption, can greatly enhance learning and creativity for humans of all ages. Developments in both "the power of play" and AI advancements share the metaphorical link of empowering learners to be creators and thinkers, not just consumers of information. We become nodes in an interactive, contextual, co-sharing, cooperative and community-oriented neural network, in a sense. Can you imagine it?

Technology, when used as a tool for active construction rather than passive consumption, can greatly enhance learning and creativity for humans of all ages.

Consumers and Creators: We Become What We Eat

As we strive to emphasize opportunities to *create* media, using a lot of principles and examples to draw from, **one of the key principles in understanding mindful media is that "consumption becomes creation."** We as leaders and designers need to create better media for others to consume so that they in turn become better creators. A corollary is that our very choice of what we consume is a creation in itself. Think of ourselves as chefs: when we create a meal, the choice of ingredients created by others is what leads to our own creation, which we consume ourselves and share with others.

We as leaders and designers need to create better media for others to consume so that they in turn become better creators.

It's a remix culture that we inhabit, and sometimes those original ingredients, and their quality, become hidden in the mix. We might not recognize that the base ingredients lack nutrients, and we might not protect ourselves from their addictive qualities. We need to be more careful. Game creators need to start caring even more about this too. Some modern games are the equivalent of junk food, and we find them compelling and addictive just as we also are left empty, jittery, and malnourished.

Again, it does not have to be this way.

We know some of the things that contribute to empty, meaningless game design that depletes our wellbeing. Let us list some of them here.

Media as Malady: Practices of Destructive Game Design That Deflate Wellbeing

Destructive digital game design techniques can have a negative impact on players' wellbeing by promoting addiction, causing frustration, and leading to unhealthy gaming habits. This is the opposite of wonder, which is predicated upon freedom. Here are some examples of how this plays out.

Destructive Digital Game Design Techniques

- **Pay-to-win Mechanics** Games that offer in-game advantages for real money can create an uneven playing field, where players are incentivized to spend money to succeed or keep up with others. This can lead to compulsive spending and financial stress.
- **Loot Boxes and Randomized Rewards** Loot boxes are virtual items that can be purchased to receive a random selection of further virtual items. The randomized nature of loot box rewards can encourage gambling-like behavior, as players may continuously purchase them in hopes of obtaining rare items.
- **Excessive Grind** Designing games that require players to engage in repetitive and time-consuming tasks to progress can lead to burnout and a sense of frustration, especially if the game does not provide sufficient rewards for the effort invested.
- **Social Pressure and FOMO (Fear of Missing Out)** Games that include limited-time events or exclusive rewards can create a sense of urgency and pressure to play, leading to excessive gaming and neglect of other life responsibilities.
- **Dark Patterns** These are design elements that trick or manipulate players into making decisions that are not in their best interest, such as spending more money or time on the game than they intended.
- **Inadequate Difficulty Balancing** Games that are too difficult or unfairly balanced can lead to frustration and stress. Conversely, games that are too easy may not provide enough challenge to keep players engaged, leading to boredom.

- **Poorly Designed In-game Economies** In-game economies that are heavily inflated or deflated can cause frustration among players, especially if they feel that their time investment does not yield adequate in-game currency or rewards.
- **Encouraging Toxic Behavior** Multiplayer games that do not have systems in place to deal with harassment, cheating, or toxic behavior can lead to a negative gaming environment, affecting players' wellbeing.
- **Lack of Clear Progression or Goals** Games that do not provide clear objectives or a sense of progression can leave players feeling aimless and dissatisfied with their gaming experience.
- **Interruptions and Time Constraints** Games that frequently interrupt gameplay with ads or impose strict time constraints can disrupt the gaming experience and cause stress.

Though these practices are decidedly negative and destructive, note that the truth of the underlying brain science can be applied in positive ways in some cases. There is nuance. For example, the concept of "disruptive game design" could involve interruptions in a positive way, employing a "Ludic Action Model," which can be interpreted as a technique that challenges players' expectations and can potentially lead to cognitive engagement. One study in particular showed that these interruptions can be positive for development and challenge.[3] However, if not carefully implemented, such disruptions could also lead to confusion and frustration, diminishing the player's wellbeing.

Similarly, discussions that highlight examples of bad game design often point out unfair difficulty levels, seemingly purposeless exercises, and illogical puzzles, which can harm the player's experience.[4] It is important for game designers to consider the potential negative impacts of their design choices on players' wellbeing and to strive for a balance that offers a challenging, engaging, and positive experience.

We have known all of this for a while. Accountability is now catching up as we expose and name the negative and accent the positive. What works best? Let us talk about some of the trademarks of games we love here, in contrast to those points earlier.

Mindful Media as Medicine: Game Strategies That Work Well for Wellbeing

These are game attributes that *play well* and are *well played*, when it comes to wonder and wellbeing, game creation strategy, and design:

Mindful Media Strategies That Work Well for Wellbeing

Transparency and Ethical Considerations

- Players are clearly informed of in-game purchases, limitations, and restrictions.
- Options are provided for players to opt out of certain features or elements.
- Tools and features are implemented that allow players to track their progress and spending, such as in-game timers and spending limits.[5]

Player-centered Design

- The player's experience is prioritized and made the primary focus of the game.
- Avoiding the use of manipulative techniques that negatively impact the player's experience.[6]

User Research and Testing

- User research and testing is conducted to ensure that the game design is engaging and enjoyable without causing harm to the players.[7]

Sandboxing

- Users are given a safe place to practice new behaviors or skills within the game environment.[8]

Clear Labeling and Messaging

- Clear labeling and messaging is implemented to inform players about data collection, in-game purchases, and other relevant aspects of the game.[9]

This list can be expanded to emphasize even more of the positive areas of agency through the "mindful media strategies" that follow. It's clear that we need to avoid the harmful techniques and strategies, but avoiding the negative is not enough. We need to emphasize the positive and produce even more data and evidence related to mindful media that show *why it works* the way it does! Not enough research has been done in this arena, in part because the playing field has not been level in the past.

These wellbeing techniques for the future of gameplay produce games that are fun, exciting, and deeply engaging—without the addiction models that left players addicted and depleted in the past. We can all enjoy a positive and engaging gaming experience while prioritizing safety, agency, and creativity—without losing what makes games *fun*! By incorporating transparency, ethical considerations, and user-centered design, game developers can mitigate the negative impact of manipulative game design techniques and promote healthy gaming environments that uplift wonder.

I had all of this in mind when thinking of my favorite game experiences and why I love them, fusing that knowledge with my appreciation for imagination, wonder, and awe and the emotional capability to empower transformational experiences.

As designers of immersive experiences of all sorts, including games, we want all humans to enter new worlds that we create and feel a sense of the elements in Figure 7.2.

Figure 7.2 Qualities of positive learning and game environments, from a player's perspective.

To see how these elements are best supported and elicited, let us look at examples from games that I enjoy, some of which you might love too, and see the mindful strategies that they employ to great effect.

Examples of Mindful Media Strategies at Play

Journey: This game focuses on exploration, cooperation, and emotional experience without competitive or stressful elements. There is heightened excitement and a dose of cooperation in the gameplay, plus elements of surprise and delight that make this game an astonishing, heart-filling emotional experience.

Animal Crossing: New Horizons: I love this game even if it irks me sometimes. It promotes relaxation, creativity, and social interaction. It operates in real time, discouraging excessive play and allowing players to set their own goals and pace. That said, the real time integrity sometimes makes me feel a certain FOMO.

Monument Valley: Known for its calming atmosphere, beautiful design, and engaging puzzles, this game aims to provide a soothing and enjoyable experience.

It Takes Two: Cooperative gaming, movement, and new environments to explore from a lilliputian-sized vantage make this game a couples' crowd-pleaser. The title says it all.

Witness: Belying its possibly ominous title, this gem of a game is set in a colorful, whimsical landscape, with puzzles to solve, hidden riddles, and brain-twisting challenges that can strike that challenge-skills balance we are all seeking. There's also no time constraint, so you can pick it up or leave it anytime you wish.

Beatsaber: It's easy to play, great exercise, and gives immediate feedback that it's working. The simplicity and correspondence with music is part of what makes it work. It follows the regular timeline-based movement, it keeps things feeling natural, lets your brain predict well, and in fact . . . you do not move toward the objects, everything comes to you, as if pulled by your magnetized body . . . all in all, a perfect way of overcoming motion sickness for the masses.

Braid: Time-manipulation mechanics and complex puzzles may promote cognitive engagement and problem-solving skills, contributing to mental wellbeing. Additionally, its artistic visual design and introspective storytelling provide an immersive and reflective experience that could contribute to emotional and psychological engagement.

Firewatch: A walking simulation, this game has no time limit, includes a lot of narration, and reveals character motivations that invite players to feel empathy and become engaged in the mystery.

World of Warcraft (WoW): A classic for a reason, this game's best parts include the competitive multiplayer aspect, which make it exciting and social. The game offers a vast and immersive virtual world that allows players to explore, interact, and problem-solve. Playing with real-life friends in games like *WoW* has been found to strengthen friendships and transfer positive gaming experiences into real life.

Kind Words: Replete with what some call "lo fi chill beats to write to," this is a unique game focused on positivity and empathy, where players share their worries and receive supportive messages from others.

These games are designed to prioritize player wellbeing by promoting positive experiences, emotional engagement, and in some cases, limiting stress and competition. While the concept of "player wellbeing" in games is subjective and can vary from person to person, these examples I give demonstrate a focus on creating a positive and emotionally enriching experience for players.

Development of the Seven ThEmes in Spatial Game Design

All of this understanding about Mindful Media and empowering, wonder-rich experience design, applied to XR, gaming, and spatial computing, went into the development of the "Seven ThEmes" I first discussed in Chapter 2. They were introduced as a template for digital wellbeing experience design. These "ThEmes" rest on top of the foundation of the Four Culture Cornerstones: Dignity, Freedom, Invention, and Agency. It's important to understand that they apply to experiences across a continuum, from physical to digital, that encompasses immersion in all forms.

Seven ThEmes of Wonder-rich Spatial Computing Design

- **Experience** (doing it)
- **Emotion and Empathy** (“in feelings” we encode)
- **Engagement** (how to invite play; distraction from pain)
- **Entrainment** (“synchrony”; aligning rhythmic elements)
- **Exercise** (“exergaming”; exertion)
- **Embodiment** (“really there”—full body + transformative)
- **Expression** (story; memory palaces)

Figure 7.3 Seven ThEmes of spatial immersive design or spatial computing game design.

Now that we have even more context for gaming interaction, let us revisit and evaluate these E’s, understanding more about their context and applications. We are spatial, dimensional beings, as this book emphasizes, and these qualities are what we keep in mind as guideposts when we take the best game design techniques, along with the four culture elements, and bring them into spatial computing environments (Figure 7.3).

Experience: The art is in the verb. Learning hinges on activation. Experience represents the psychological phenomenon that occurs when a player’s consciousness is temporarily expanded into areas of the unconscious, akin to hypnosis or lucid dreaming states.[10] It involves a process that draws people in and transports the player into an alternative world where techniques are used to make them feel more like the character they are playing. This is achieved through a combination of good game flow, fully developed storylines, and extreme realism made possible by the quality of graphics and interaction design.

Emotion and Empathy: All meaningful experience is encoded in emotion. Emotions give it resonance. When we feel what others are feeling in an experience, we call this “empathy,” Greek for “in-feeling.” Deeper than sympathy, empathy is to feel with someone. This drives perspective taking, better understanding, and compassionate action. Emotions themselves allow us to feel, and empathic emotion lets us use our emotional range in immersive spatial design to inhabit others' perspectives, thereby transferring that learning to improve the human condition at a societal "we" level. While we could separate "Emotion" and "Empathy" into two separate "Es" here, I choose to keep them connected because they are so linked. Note that not all emotions are empathetic ones—a vast spectrum of emotions can be evoked by a spatial experience, and they can be deep, resonant, and transformative. Wonder and awe are two of these emotions.

Engagement: It's not enough to watch, and it's not enough to participate in an experience. The level of engagement is what drives the experience. It's the interaction and the invitation. When you engage with something, you use it and thus become part of it as it becomes part of you. You are invested. Engagement is that coupling of intention and attention. In this way, it is energetic. Ideally, it's power-giving, em-power-ing. Mindful engagement design in spatial employs the best of what we know about psychology, science, and health, and it can serve to help distract someone from pain and to rewire their habits. With great power comes great responsibility.

Entrainment: The "internal flow" of the experience, which a player or participant will follow, is entrainment. It refers to an internal rhythm, the flow of tension and relief within the interactive content, which can mindfully create a hypnotic and engaging experience. This concept is essential for understanding and creating interactive content, as it helps maintain a balance between tension (e.g., active questions and user actions) and relief (e.g., passive transfer of knowledge) to ensure an effective full experience. If it is mindful and compelling enough, we will follow, sparked not only by our curiosity, but also by our biorhythms, which are naturally designed to mirror and mimic. When used mindfully, entrainment can integrate rhythmic and audio-visual elements in response to sensory biofeedback integrations in a way that adds to that overall flow state and player immersion. For example, the classic game *Pac-Man* is cited as an example of entrainment, where the ebb and flow of tension, coupled with audio-visual cues, creates a hypnotic and engaging experience for players.[11] Furthermore, research has explored the impact of ambient sounds on player immersion in video games, demonstrating how audio elements can contribute to player flow state.[12] Entrainment makes a difference.

Exercise: A hallmark of mindful media, movement and exercise have been scientifically shown to be beneficial for wellness, for learning, and for overall body and brain health. "Exergaming" has a tremendous market driving it in the immersive gaming world. The beauty of immersive media in spatial computing is that it naturally lends itself to physicality by being three dimensional–and, fun! I recently participated in a study run by Columbia University associated with movement and a series of podcasts called "The Body Electric," to be discussed further coming up. Not only are we primed and eager to move, chairs and couches are a relatively modern invention, and our blood sugar benefits from regular movement and exercise.

Embodiment: This aspect of design has many facets and considerations. Some would say that all of spatial technology is about embodiment. Basically, we are 3D creatures designed and built to feel our experiences and invent a frame for the world through our sensory inputs. Embodiment, as the word implies, is the inhabiting of this shell, this vessel, that allows us to make the connections that provide us with a constructive understanding of the world. Because we are *embodied*, this way of knowing that we have is relational to our place in it—we are not passive viewers. In spatial computing, the notion of embodiment, then, is deeper philosophically than simply adopting an avatar. It has to do with sensemaking and deeper ontological meaning. I'll talk a bit more about embodiment later with examples; it's a provocative topic with a lot of extensions in VR design.

Expression: One of the most powerful opportunities spatial computing and immersive game design provides is the chance to create moving representations of . . . *anything that moves us*! It's a realm of creativity for each player and participant. It's a space and place that allows imagination to come alive and become less abstract, more concrete, as it takes the shape and form that we want to create. Because of the ease of modern spatial computing technology, creators do not have to be experienced programmers—they can use their passion and ingenuity to drive expression, using creator-friendly tools and interfaces. When we express ourselves, wellbeing receives a natural boost too, as evidence shows.

Engaging in creative arts and self-expression has been shown to have significant positive implications for mental, physical, and emotional wellbeing. Imaginative expression can improve wellbeing by helping individuals understand themselves, shift perspectives, and reinforce positive behaviors.[13] It provides a powerful outlet for emotional release, fosters emotional awareness and growth, and serves as a coping mechanism, ultimately promoting mindfulness and reducing stress.[14] Physiologically, engaging in creative activities such as the arts or storytelling has been found to reduce blood pressure, bolster the immune system, improve brain cognition, and fight inflammation. Furthermore, creative expression through the arts can build self-confidence, foster community belonging, and support pro-health behaviors, contributing to improved wellbeing.[15] Because digital immersive tools give so much creative freedom, it's a natural fit for a powerful form of mindful media that can spread wonder and delight.

Why Embodiment, Wellbeing, and Somatic Practices are Core to Game Design, Spatial Computing, and XR

I design experiences that center on human physical experience. We are physical creatures, and immersive technology can be thought of as layers on top of the physical. The physical primary layer is the most important one, in many senses. Understanding this impacts your life as a leader, as a learning designer, as a game creator, and as a consumer of any digitally mediated experience. Let us investigate why this is so critical to our future of work and play.

Embodiment in the context of spatial technology, immersive design, and gaming refers to the integration and representation of physical presence and interaction within digital environments. This concept is crucial for creating immersive experiences in VR, augmented reality (AR), and other digital platforms, where users can feel as though they are physically part of a virtual world. Embodiment is achieved through technologies that enable users to see, move, and interact with the digital environment in a way that mirrors real-life movements and interactions. This can significantly enhance the user's engagement, presence, and overall experience within the digital space.

Imagine, for example, a VR art gallery experience. Users wear VR headsets that transport them into a digital space mirroring an art gallery. Motion tracking technology integrated with the VR system mirrors users' physical movements, allowing them to navigate the virtual space, interact with artworks, and experience immersive visual and spatial feedback. This embodiment is achieved through physical presence and interaction, visual and spatial immersion, and interactivity, creating an engaging experience that bridges the gap between physical and digital spaces, enhancing overall user experience (UX) and engagement with digital content.

Embodiment and Wellbeing in General

Embodiment in digital environments, particularly through somatic practices, has implications for wellbeing. Somatic psychology emphasizes the importance of body awareness as a tool for healing, suggesting that how we move and feel influences our perceptions of ourselves and the world.[16] This bidirectional feedback loop between movement and perception is central to embodiment. In therapeutic contexts, focusing on embodied experiences can help individuals reconnect with their sensations as part of the healing process, offering pathways beyond traditional verbal processing.[17] This approach can improve emotional regulation, decision-making, and social interactions by fostering a deeper connection with one's internal states and responses.

Embodiment's Implications for Digital Wellbeing and Wonder

The concept of embodiment relates to digital wellbeing by highlighting the importance of creating digital experiences that support and enhance users' physical and psychological health. In immersive design and gaming, this means developing interfaces and interactions that promote natural movements, reduce cognitive overload, and encourage positive emotional experiences. For instance, VR technologies that enable embodiment can lead to improved skills such as spatial perception, memory, and navigation, which are crucial for various cognitive tasks and everyday activities.[18] Moreover, experiences of embodiment in VR have been shown to influence behavior, attitudes, and cognition, suggesting that immersive technologies can have profound effects on users' wellbeing.[19] Using these understandings and incorporating wonder into the experiences can expand our awareness, our concept of "self," and enhance imaginative applications, among other benefits.

Spatial Design Practices Integrating Somatics

Somatic means "of the body" or "relating to the body" particularly in contrast to the mind or mental processes. Design practices that integrate somatic practices into digital environments focus on creating experiences that are mindful of the user's physical and emotional states. It helps to remember, as I emphasize throughout this book, that we are physical-first creatures adding layers of digital on top of our primary biological embodied sensibility. This understanding urges me, as a creator and experience designer, to design interactions that are not only intuitive but also responsive to the user's bodily movements and sensations.

For example, VR experiences that incorporate body tracking and haptic feedback can enhance the sense of embodiment by providing users with direct and meaningful ways to interact with the virtual world.[20] These design practices can lead to more engaging and beneficial digital experiences, promoting wellbeing by encouraging users to remain connected with their physical sensations and emotional states even while immersed in digital environments.

Embodiment in spatial technology, immersive design, and gaming is a multifaceted concept that plays a crucial role in enhancing UX, spatial skills, and overall wellbeing. By integrating somatic practices into design, creators can foster deeper connections between users and digital environments, promoting healthier and more meaningful interactions with technology. Leaders now know what to look for and prioritize in a digital experience. The benefits for you and your teams when applying spatial learning and collaboration in the workplace are exponential. It can impact all human-centered relationships, having tremendous payoffs in every area, from health to engagement to innovation and profitability.

Somatic Practices That Can Enhance Wellbeing Every Day

Somatic practices encompass a variety of modalities aimed at deepening the mind-body connection. These practices can be integrated into experience design in various ways, and it's important to note that they do not have to be for people who self-identify as dancers, athletes, or performers. These are for everyone, and knowing more about them can enhance our everyday life, providing even more onramps and access to that wonder-infused plane of wellbeing we talk about.

Dance: Somatic practices in immersive mindful media movement and dance education provide humans with a deeper understanding of their body's movements and sensations, fostering an embodied approach to movement and performance.[21]

Yoga and Tai Chi: These practices involve integrated somatic postures and movements to develop harmonious body awareness and movement, which can be applied to product design and human-computer interaction.[22]

Mind-Body Medicine: Mind-body medicine is an approach that recognizes the interconnection between the mind, body, and behavior, utilizing somatic practices like mindfulness, breathwork, and movement to promote self-awareness, self-regulation, and overall wellbeing.[23]

Mindfulness and Movement Practices: These simple practices can be applied to awaken body awareness as a tool for healing in somatic psychology. We can use somatic education to improve movement and enhance human functioning through increased self-awareness and improved coordination.

There are many more specific somatic movement practices and methodologies. By integrating different types of somatic practices into daily routines and immersive design, we can develop products and experiences that are more attuned to the user's bodily sensations, emotions, and overall wellbeing. These practices are complements, not substitutes, for other types of supportive therapies that people might need. Increasing our awareness of somatic principles, in my view, serves to empower and enhance our "primary layer" connection with our physicality before we add layers of digital technology. We also become more connected and aware internally of the power of our mind.

"Somatic" is one of the seven S's of a mindful metaverse, a framework I developed to help increase awareness of all of the factors that influence effective connection and collaboration in immersive imaginative environments. We map those seven S's coming up in Chapter 8.

Emotion in Games

Development of the seven E's and overall findings are reinforced by views of top emotional design experts in the field, including Dr. Celia Hodent, a game UX expert and former director of user experience at Epic Games, known for her work in applying UX and cognitive science in product development. She has called empathy the most important skill that we can develop "to create better games, better teams, and a better world."

Hodent's work emphasizes the application of cognitive psychology and neuroscience to create games that are not only engaging but also considerate of the players' wellbeing, experience, and intrinsic motivation. She advises integrating principles of perception, attention, and memory into the development process, emphasizing the importance of understanding the limitations of the human brain in these areas as well as the emotional and motivational aspects of the player experience.[24] By doing so, game developers can create a compelling UX that takes into account the cognitive and emotional capabilities of the players.

For example, in *The Legend of Zelda* series, the gameplay requires increasing mastery to progress, which aligns with the principle of satisfying the basic psychological need for competence.

In *Minecraft*, players can experiment with the game environment in a creative way, offering meaningful choices and opportunities for self-expression, which aligns with the principles of autonomy and intrinsic motivation.

These examples demonstrate how games can be designed to satisfy basic psychological needs for competence, autonomy, and relatedness, thereby enhancing player engagement and experience.[25]

Culture in Games

Another critical consideration that informs mindful immersive design involves culture, as Kate Edwards knows well. Edwards is a prominent figure in the video game industry, known for her work in geocultural content advice to game companies. She is chief executive officer of Geogrify, chief experience officer and cofounder of SetJetters, and has held various leadership roles, including serving as the executive director of the International Game Developers Association (IGDA).

Edwards has been an outspoken advocate for diversity, inclusion, and better working conditions in the gaming industry. She has also been involved in addressing issues such as crunch time (forcing developers to work unpaid overtime) and promoting mental health support for game developers. In July 2023, she gave a talk at the Games for Change annual conference titled "How Our Values Impact Our Games' Cultural Influence." Edwards pointed out the importance of recognizing our own blind spots and assumptions as creators. She said that what we envision, what we build, and how others perceive that game world have to do with a culturalization and compatibility with a local world view. We need to recognize that "all games have a cultural dimension."

If mindful immersive media is also about the larger values and contexts at play, and our broader goals to promote positive change in society, we must recognize that "content carries culture, culture contains values, and values catalyze change." Edwards reinforces the power of media, reminding creators that "to deny or ignore that your games and their carried values influence mindshare is to reject the very power of your medium."

Everything is by intention, and our choices about what to prioritize when it comes to wellbeing are making a case for our intentionality. It's a call to action to consider and take seriously our choices. Our four Culture Cornerstones of Dignity, Freedom, Invention, and Agency are further bolstered by this empowered awareness and action in game design that take into account cultural context.

Mindful Game Creation

When we build games, it's important to think about the culture and the overall goals of the game. Who is it intended for, and what are the intended outcomes? What are we building into the story that will deliver that message as an interactive experience rather than a lesson? Are we inviting a player to feel it, and to have a sense of agency and empowerment along the way? Where are the elements of wonder and delight? So many considerations go into our mindful game creation.

There is an interplay between rapidly advancing technology and the slower evolution of critical institutions. One solution that reflects wonder and wellbeing involves adaptability, using more agile frameworks that can respond to and grow with these technological changes.

Alan Gershenfeld has a unique perspective on gaming, technology, and their potential for positive impact, thanks to his leadership at E-Line Media. His portfolio includes *Never Alone* and *Beyond Blue,* games that have broken the mold by combining entertainment with education and social consciousness. Before E-Line, his influence at Activision Studios saw him guiding the development of classics such as *Civilization: Call to Power* and *Tony Hawk Skateboarding*. His commitment to leveraging interactive media for good extends to roles with Games for Change and contributions to significant gaming initiatives, which include working with foundations and governmental bodies.

When discussing the essence of game design with Gershenfeld, we recognized games as tools for social good:

> *"Games have enormous potential for positive social impact because they let you step into different roles, take on challenges, fail safely, get feedback, and move closer to goals you are invested in (both individually and collaboratively). They are interactive, participatory, learn forward, and give players* agency. *If this agency can cross over into the real world it can be especially powerful. Games also provide valuable community and affinity spaces—which is particularly important for marginalized youth. On the other side, there is still a lot of toxicity in gaming communities, some predatory business models (especially in mobile), inappropriate content for different age groups, excessive time playing—all of which the industry needs to be more effective at helping to address."*

As for the creative process, Gershenfeld shares that E-Line Media prioritizes commercial games "with meaningful themes and authentic voices. We have a passion for exploring under-represented cultures and voices through an 'inclusive development' process, exploring the awe and wonder of nature and the planet as well the big social issues of the day where they organically align with fun gameplay." They show a commitment to these values as they continue to develop new titles that cater to players' wellbeing, from those that foster collaboration and connection to experiences that facilitate intergenerational play or open doors to new creative ideas.

Gershenfeld reflects on the role of awe in games designed to effect change. When I asked him where awe and wonder fit into the development process, he told me, "Our games are often inspired by amazing real world partners—whether Alaska Native elders, writers or storytellers, marine scientists/aquanauts, investigative journalists, scientists etc. We do not make games *about* these folks, we make games collaboratively *with* these folks."

Thinking about the awe that we can find in another person and their story is inspiring. We can find cultural custodians, engaging not just as communicators but as co-creators. This collaborative process does not just craft games—it forges pathways to new understandings and shared experiences. These are foundational elements of the future of mindful game creation.

The Significance of Fun in Wellbeing Game Design

Games have a profound ability to positively influence our wellbeing. Full stop. We know this innately, from our childhoods spent inventing and playing games. And they are not limited to childhood years. Everything about being human, throughout the lifespan, is celebrated in games, and we have used games throughout human history as ways to learn, innovate, empathize, entertain, share cultural values, and experiment. We use games to connect, inside and out. The legacy of Papert supports this: Projects, Passion, Peers, and *Play* are embedded in the best experiences. Games, through their *play*, aka, *fun* design, can cater to our innate desire for learning, pattern recognition, and problem-solving. This aligns with broader views that "fun" in games arises from encountering and overcoming challenges, which in turn stimulates our brains in beneficial ways.

We use games to connect, inside and out.

Raph Koster, former chief creative officer for Sony Online Entertainment, has shared various insights on the intersection of wellbeing and game design, emphasizing the power of games to contribute positively to players' lives. Known for his contributions as the lead designer for titles including *Ultima Online* and *Star Wars Galaxies,* Koster's 2004 book *A Theory of Fun for Game Design* explores the concept of fun in games and why it's the most vital element in any game.[26] He says fun arises from encountering and overcoming challenges, which stimulates our brains through learning, pattern recognition, and problem-solving. His theory is widely respected and used in various university-level programs on game design and by professionals in gamification, education, and interaction design.

Wellbeing and Game Design

What exactly is wellbeing in game design? Since I view it through a lens of wonder, I would say that wellbeing in all design, including game design, is a state of open possibility, uplifting and expanding our senses. It is a capacity to lead to greater happiness and a richer, fuller experience, connected to a moment-by-moment appreciation for the unfolding of the game. Thus, wellbeing in game design also has to do with a fuller sense of presence, to operate on that higher plane.

Koster defines wellbeing in game design as the capacity of games to create greater happiness in the world. He believes that games can teach us to focus on what really matters and encourage systems thinking.[27] Koster sees games as a medium that can change how we think and perceive the world, potentially leading to positive changes in behavior and thought processes.[28]

In his discussions, Koster has highlighted the concept of "serious games"—games designed with a purpose beyond entertainment, such as education, training, or health improvement. This falls right in line with Seymour Papert's views about hard play. Koster sees immense potential in these types of serious games to redefine our perception of fun and contribute significantly to our wellbeing. For instance, he imagines games that could help players build immunity to diseases or learn new forms of medical intervention and diagnosis, suggesting a future where games play a crucial role in development of health and education.[29]

This is where the importance of user-created content and sandbox design can make a difference, already identified by us as signatures of Mindful Media. Our methods allow players to express themselves creatively and engage with the game on a deeper level. This approach can be seen in games like *The Sims* and *Minecraft,* among many others, where players are empowered with significant creativity and freedom to invent and explore.

Through serious games and designs that encourage learning, creativity, and social interaction, games can go beyond mere entertainment to become tools for education, health, and personal growth. Koster's "Theory of Fun" is a major contribution to wellbeing knowledge that expands on Papert's legacy and beliefs. Because Koster defines wellbeing in game design as the ability of games to contribute to happiness and personal growth, he shines a light on wonder-infused digital wellbeing's importance and priority in a field where it is sometimes overlooked.

This points toward a future of integration where leaders, designers, creators, and players can establish a mindful media culture, celebrating games that enable meaningful values-centered learning, creativity, and social engagement.

Understanding Spatial Computing and Its Impact on Wellbeing

Spatial computing represents a significant leap in how we interact with digital environments, blending the physical and virtual worlds in a way that feels natural and intuitive. This technology has profound implications for wellbeing. Now we, as leaders, can also see their potential for dynamically changing the nature of personal and professional life and whole-life thriving.

Revisiting the Special of Spatial

Spatial computing is special because it allows for a seamless integration of digital content into our physical world, making interactions with virtual objects and environments feel as natural as those in the real world. In this book, we have been easing into the topic of spatial computing by approaching it in context of our overall wellbeing. It is a layer and an extension of our physicality and our full-bodied sensory experience. This is why we approach "digital wellbeing" as one full integrated term, not a hyphenated, segmented topic. It's intentionally holistic.

When we use spatial computing, we are working with advanced technologies that run the gamut of XR, including AR, VR, and mixed reality (MR), which together create a fully immersive digital environment. Devices like the Apple Vision Pro, Microsoft HoloLens, and Meta Quest Pro use spatial mapping, eye-tracking, motion sensors, and hand-gesture detection to enable users to interact with digital content in a way that mirrors real-world actions.[30]

Spatial Applications in Wonder and Wellbeing

We've already shown deeply, through our explorations of facets of wonder and XR applications, all of the links. Spatial computing has the potential to revolutionize the entire wellbeing industry by offering immersive experiences that can aid in relaxation, meditation, physical therapy, cognitive rehabilitation, and more. For instance, immersive wellness programs leverage this technology to provide holistic experiences that help recharge the mind and body. These destinations offer a range of activities and environments designed to promote mental and physical health, from serene landscapes for meditation to interactive exercises for physical therapy.[31] With a new range of multimedia sensory-rich integrations that tap our imagination, we can think of these applications as vehicles to animate wonder and bring us in closer connection with ourselves, each other, and our worlds.

The Role of AI in Spatial Design

AI plays a crucial role in the design and development of spatial computing experiences, particularly in making these experiences more intuitive, personalized, and effective. Generative AI, for example, can be used to create dynamic and responsive virtual environments that adapt to the user's actions and preferences. This allows for the creation of interfaceless systems, where interactions are based on natural gestures and movements rather than traditional interfaces like keyboards or controllers. AI can also automate the testing and optimization of these environments, ensuring they meet the desired usability and effectiveness criteria.[32]

AI's involvement extends to enhancing the UX in spatial contexts by improving automation, functionality, aesthetics, and the overall design process. This includes leveraging AI for more efficient prototyping, testing, and iteration of spatial computing applications, ensuring they are both user-friendly and aligned with our wellbeing goals.[33] As ever, intention can drive attention, and then everything is a "fit for flow" model, as I like to say, because we fit our tech integration to optimize for our best flow experience. It's not a protocol as much as a possibility.

Challenges and Future Directions

Despite its potential, spatial computing faces real challenges, including the cost of hardware and the need for significant investment in research and development. Additionally, integrating spatial computing into daily routines and business operations requires overcoming technical barriers and cultural acceptance. Everything is about consideration. The benefits of spatial computing, such as enhanced learning, training, and therapeutic applications, outweigh these initial hurdles. Also, we need to continually emphasize that the way of wonder-rich digital wellbeing is not an either/or proposition of physical and spatial. It's about creating layers of intention-driven technology and honoring our physical reality first.

As spatial computing technology continues to evolve, we can expect an increasing number of applications in wellbeing, across all different sectors. The integration of AI has the potential to further enhance the design and effectiveness of these applications, making immersive wellbeing experiences more accessible and impactful. That said, it's all about the intention and the quality. Doing something mindfully takes a quality of care that is uniquely human and also an attentiveness to the training models, the ethics, and attunement with overall goals.

This is why I've been intentional about my own journey and ways of mindfully using AI. What is most elegant? What feels as if it is filled with wonder and inspires awe in us? We need to feel this sensibility along the path, marveling at ourselves and each other in the process. As Jaron Lanier says, we need data dignity, which I couple with data diligence. The AI itself is nothing to worship. It's my hope that this process will invite us to refine our relationships with each other, with ourselves, and with nature.

Spatial computing, with its ability to merge the physical and virtual worlds, offers many opportunities for enhancing wellbeing. The involvement of AI in the design process can help efficiently guide intuitive, personalized, and effective immersive experiences. Despite facing challenges, the future of spatial computing in wellbeing looks promising, with ongoing advancements certain to broaden its applications and accessibility. Let us stay conscious of our "why" intentionality along the way.

Benefits of Using Spatial Computing for Wellbeing

- **Enhanced Immersion.** By creating a more natural and intuitive interaction with digital environments, spatial computing can help users achieve deeper states of meditation and relaxation.[34]
- **Exergaming.** Immersive workouts can combine psychological excitement with physical exertion, potentially increasing the likelihood of users sticking to their fitness routines. Approaches to exercise can use AR apps to track movement, suggest routes, and allow users to share progress, potentially making fitness activities more enjoyable and engaging.[35]
- **Improved Diagnostics and Training.** AR and VR can enhance the accuracy of medical diagnostics and provide more effective training for medical professionals.[36]
- **Accessibility.** Technologies like smartglasses and virtual home care nurses can make health care more accessible to people with disabilities or those who require constant monitoring.[37]
- **Mental Health Support.** Platforms can use immersive technology to provide scalable access to mental health experiences, which can

be particularly beneficial in addressing the global mental health crisis.[38]

- **Personalization.** Spatial computing can offer personalized experiences tailored to individual needs, often helpful for effective wellness and therapeutic interventions.[39]
- **Data-driven Feedback.** Spatial computing devices can provide users with detailed data about their athletic experiences, for example, such as distance covered or heart rate, which can be motivating and help track progress.

The Future of Spatial Computing, Imagination, and Wellbeing

It's clear that spatial computing, imagination, and wellbeing are meant for each other because they are expansive by nature. Studies have been showing the benefits of many forms of XR applications in areas of health and wellbeing. I highlight much of this ongoing research on my website and continue to apply my own research and work to broaden the understanding of the immense potential of immersive experiences to animate wonder and awe, help us manage stress and anxiety, and increase a sense of belonging and connection instead of loneliness.

In areas including mental and physical health, media and medicine, spatial computing is emerging as a transformative technology. Through our explorations, we learn that immersive media—including spatial computing, which encompasses gaming and XR—can be medicinal and more, amplifying imagination and enhancing wellbeing. It's all about the design and execution. Through immersive, engaging, and personalized interactions with digital content, experiences become part of us.

The signature aspect that distinguishes this medium is its ability to evoke feelings of wonder, awe, and flow states, combined with layers that include music and entrainment. Now is our chance to add to this chapter's "bead on the necklace" with even more augmented wonder, looking toward the possibility of a fully integrated digitally mediated spatialized future that builds on our essential humanity.

Personal Reflections

1. What were your favorite games to play growing up? Did you feel "immersed" in them, and did they bring about a sense of wonder? Why do you think they appealed to you?
2. How do you get a sense for when a game is designed with your wellbeing in mind?
3. Have you ever played a game that was fully spatial and immersive? Were you in a headset? How did it feel, before, after, and during, as you embarked on a "Hero's Journey" crossing thresholds?
4. How do the Seven E's affect your understanding and approach to spatial computing? Was anything surprising or especially intriguing about those elements?
5. Seeing the benefits of spatial computing for wellbeing, what are you looking to explore more deeply, and in what medium?

Reflections for Leaders

1. Who has played the role of a mentor in your life as you explored models that stretched your thinking and ways of being? How might your experience have followed a "SCUBA framework" as outlined in this chapter?
2. How have you played the role of a mentor and guide for others on a Hero's Journey? How does the "before" and "after" of those experiences crossing thresholds make a difference?
3. With Papert's legacy in mind, how do his views about constructivism affect your approach to AI integrated into work practices?
4. Does your working life encourage and practice forms of wonder and play? If yes, how? If not, why not?
5. Now, looking at different theories of game design and wellbeing, how might we incorporate more mindful media design and gaming practices that empower imagination into our leadership and work culture?

Extensions for Further Exploration

I am interested in extensions and future explorations that investigate effective use and inducement of neurotransmitters and rewards systems, with more of a neurochemical analysis. I would also like to engage in longitudinal studies as well as look into even more game analysis and studies of players' feedback. I'm also deeply interested in using these understandings to empower younger generations and to reform and expand our work culture, offering even more ways to deeply connect with ourselves and each other, and to infuse imagination and wonder into the workplace, which is critical in this age of AI. We can do this in ways that are not addictive and not destructive. It's time for the anxiety-inducing algorithms of the past to be identified and rejected, and time for a new data dignified system to be in place that prioritizes human freedom when it comes to digital wellbeing. This will make a difference for the future of humanity. There is so much at stake. In the next chapter, we look even more deeply at the worlds of XR, stretching that conversation even further and addressing the future that we now have a chance to reshape. Let us make an impact.

8 Wellbeing Through Immersive Imagination

The Future of Engaging Spatial Computing, by Design

In this book, we've explored many "beads on the necklace" of digital wellbeing, recontextualizing concepts to fit our lives in a way that enhances thriving. This chapter is all about spatial technology and its ability to amplify imagination, using that immersive imagination to increase wellbeing in all dimensions of life. I want to make these methods accessible for everyone, focusing more on experience and less on hardware, which is constantly evolving.

You've come to know more about my history and what inspires me to help create experiences across a continuum of physical and digital media, applying them to education, business, health, personal life, and more. All of these forms and groups I work with have the opportunity to approach experiences using a lens of wonder. In this way, all experiences become ones that are transformative. We as participants become active agents in these engagements that become transcendent as our imagination broadens our awareness and our skills as leaders and connectors. It's all in the design and the priorities, with intention.

This type of mindset, and skill building, is changing the future of everything.

Words in this book are their own medium, a translation across the divides of time and space, from me to you. What I write in this "now" moment will reach you in your "now" moment and become entirely different. Each person's reading of this book will be different, and your experience of it will be unique each time you come to it because you are constantly changing and your context will be distinct with each reading. This is true with all forms of art. It's never a static encounter. Each time, it is important to arrive and show up as we are, and to take what is useful to help us rise, using the imagination inside of us.

I say this because we're about to talk about spatial computing even more deeply in this chapter, a topic blocked or complicated by the names we give things and the conventions we associate with those names. We form expectations and predict our way into our own "reality," which is different from allowing the source to emerge through awareness. This different approach has to do with an open way of being. Reflecting that openmindedness, skateboarding legend and imagination pioneer Rodney Mullen offers us an Interlude here:

On Imagination and Extended Reality, by Rodney Mullen

I'm honored to share a few ideas here that frame Caitlin Krause's book about imagination and holistic flow. Using a metaphor of a web interface, I'll rattle off ideas in the form of browser tabs, for you to connect as you will.

Skateboarding has mostly defined my life since I was a kid. It's given me all I have and taken me further than I had the courage to dream. In the way, say, artists connect with their medium, and stunt drivers are "at one" with their cars, so it is for us with our boards.

Tab 1: if you don't know the *Rubber Hand* experiment, take five minutes to YouTube it: our brains not only have the power, but a comedic propensity to integrate the inanimate into our sense of self with rippling implications—into not just how or what we perceive, but to allow that inanimate to become part of the models through which we conceive. It happens so subtly that it flies under the cognitive radar, blurring the "lines" of self/nonself.

Hold that thought.

Tab 2: A kid learns to play the piano by mapping movement to sound via rules of harmony and the architecture of the piano; likewise, skaters map movement via rules and structures to do tricks (with such precision) that could otherwise land them in the hospital for being a fraction of a second or inch off. Also, we infuse parts of ourselves into ideas that turn to actions, evincing the same from others who made the trick what it was to us—it's a cascade of learning. Next, as we ingrain these movements over thousands of hours, the physics of the environment in which we learned them (pools versus street, etc.) bakes into the movements so inextricably that you can't get 'em out, even over decades; worse, like an accent, we don't even realize it until we are out of our native domain.

Tab 3: Our brains infuse and integrate sense data and emotion into reason so deeply that it is integral to our ideation and decision-making, even undergirding a kind of auxiliary language, which athletes, dancers, musicians, etc., develop by repetitive practice. This encodes the ones and zeros of whatever kinesthetic code until the trick becomes *executive*. The analytic mind treads lightly as it calls the trick forth, rhythmically counting off somatic markers we use as guiding tags to thread the movements together. As the trick becomes mediated by fewer tags, "thinking" gives way to the faster, quieter mind.

In the way quantum behavior undergirds superconductivity, this inner language of synaptic idiom is to *flow*, which becomes integral to the very models through which we think, envision, and imagine.

Add this to Tab 2: The baked-in stuff factors so much that entire life's works are based upon it—mine, in particular. Now, pull-in Tab 1, where the rubber handing of the inanimate becomes so intertwined with not only the language, but the models and feeling through which we envision and stack our conceptions that it's not only impossible to differentiate, but it is inextricably enmeshed into rational thought, tunneling into biases, and even into our sense of self.

Over the last couple of decades, technology has deluged this generation to bring about changes faster than any experienced in human history. As it was with Oppenheimer or Dr. Frankenstein, we're tapping into forces with the capacity for great benefit or powerful destruction. Warning signs have emerged, with not-so-secret machinations to manipulate behavior over the aggregate that can determine the fate of nations, perhaps loudly with social media–influenced revolutions that continually refine, for better and worse.

Yet, we are now at the precipice of a new potential to enrich our interior lives through the emergent power of extended reality (XR), spatial computing, and other digital technology to augment the world, as it is, and integrate the virtual so seamlessly into the models through which we think and engage that we become one with it. Through these potentially wonderful, awe-inspired media, we can connect with others in ways that plumb the depths of language, of art, of being itself. We are *fearfully and wonderfully made.* Although we float on calm seas of human potential, few of us doubt a mother can lift a car to save her pinned child or a crash survivor saying how time slowed down; these are mundane reminders of how the will can tap into deep reservoirs of staggering potential, latent in us all.

As Caitlin invites us to take a journey of imagination through this book, it could be a chance to remind ourselves that the process will involve unlearning and openness as we explore dynamics that are more fluid than static, more integrated than separate. It's up to us to navigate the perils and embrace what is to come with wisdom and humility.

—Rodney Mullen, February 2024

On Presence and Showing Up with XR

With grace and purpose, let's segue and start with the topic of presence. I was part of an EdX program offered at MIT called "Presencing: In Business, Society, and Self." The course calls itself a "ULab"; going deeply down and inward to then propel outward and impact the world in a meaningful way, with presence. The process requires awareness and patience. The course uses a schema of "Open Mind, Open Heart, Open Will," encouraging participants to suspend judgment, view with compassion, and use openness to tune awareness of an emergent future, meeting the world where it's at. I take time to mention theories (learn about them more at presencinginstitute.org) because I see openness and fuller presence as key missing elements from the narrative around us with digital wellbeing and mindful, imaginative tech adoption, especially with all of the cynicism and forms of destructive online communication happening today. I am inviting a shift in engagement, open to it at all levels. The possibilities with immersive spatial technology are too ripe and amazing not to apply these principles and intentional philosophy.

Let's start by naming some of the ambiguity so we can clarify it. Several factors are at play that cause confusion and distraction in immersive tech. Many people first think of XR and spatial computing in terms of the hardware. They treat the headset as if it's a hat, or better yet, a shoe. How much cushion does it have? Is there enough room for movement? What is the weight? How do I look in it? What do you think of the design? Their next stage of evaluation will ask: *What can I do with it? How does it make my life easier?*

While physical ergonomics and comfort factor are important, the transformative evaluation stage I'm interested in brings a series of important questions: *How does it change my way of thinking? How do I feel when I'm using the headset*? and just as importantly, *How do I feel right after I use it, and throughout the remaining part of the day?* And, ultimately, *Who am I* as I use these integrated devices for work, for recreation, for social enjoyment and personal wellbeing? Who am I, and who am I becoming? Who are we becoming as we engage with these experiences together? Are we growing and stretching in our self-awareness? Do we have sovereignty and integrity as we pass through these thresholds and come out the other side? How are we each involved in shaping our multilayered journeys?

Virtual reality and spatial computing have great and imminent potential to play a major role in shaping the next phase of human development as a species. We are already entering that phase, laying the groundwork and some of the "rules of engagement" in a sense. Here is a chance to talk about it.

Rules of Engagement

Let's think of "rules of engagement" before "tools for engagement." These rules, or standards and guidelines, will guide the choices we make about lifestyle design and hardware. We've already determined that all headsets are not created equal, and they shouldn't be approached in an "X versus Y" sort of way of either/or comparisons. Which hardware is better? It's an evolving landscape of constant refinement. By entering the HMD (head mounted device) market, stakeholders have signaled that they believe in the longevity of the industry, and products are subject to iterations and improvements. We should be focusing on what they provide in *experiences*, along with the other E's in our ThEmes. Certain hardware and platforms are optimal for certain use cases, not for others. This understanding lets consumers be part of the conversation as companies iterate to meet needs of what will eventually be a part of our ways of working and sharing. Extended applications are here

Extended applications are here to stay, and as spatial creatures ourselves, it makes sense that we have an "extended mind" and many ways of sharing and collaborating that are in a 3D world.

to stay, and as spatial creatures ourselves, it makes sense that we have an "extended mind" and many ways of sharing and collaborating that are in a 3D world.

"Engagement" came up in our Seven ThEmes framework and descriptions in the last chapter, and it's a top consideration. When we engage with something, we are interacting with it, using all of our openness and curiosity to really see it, and to allow it to sense us, in a way. It's like having a great dance partner. There's an adequate and palpable amount of tension in the interaction if there's engagement, a push and a pull. This is what allows the dance to happen, because there's energetic cosensing and direction at play. You don't engage passively. You also don't know exactly what's about to happen, so that's part of the fun. If it's engaging enough, surprise and delight are infused into the architecture and the interaction design.

Ideally, this sort of Engagement evolves to incorporate Entrainment, a natural tension and pacing that leads toward flow state. We will talk about flow state more deeply in this chapter, and give some examples of immersive spatial computing places that are actively using flow state principles to create meaningful experiences that stretch us in new ways.

The Metaverse: Bringing People Together

The "metaverse," as we've talked about, can be defined simply as a digital world that's shared, spatialized, and interactive. Call the "interactive" part "social" instead and you have the first three S's of metaverse: ***shared, spatialized, and social***. This is just the beginning of exploring the wonder.

Experiences with astonishing, awe-inspiring spatial worlds and shared real-time events can be extraordinary in many ways. When people are together in a shared environment of a metaverse, something remarkable happens mentally, physically, and emotionally. There's a powerful ritual taking place as humans opt in to a shared experience. The key is connection, presence, agency, and psychological safety.

We talked in Chapter 7 about the XR Hero's Journey and the SCUBA framework that involves the "before" priming and "after" application and transfer of understanding. It's about the entry and exit as much as the heart of the experience itself. It's also about a collective sharing of meaning. Someone taking part in an experience is asked to self-reflect and to genuinely show up as part of the group. They matter.

Shared immersive imagination in a collective "metaverse" serves many purposes, for individuals and groups:

We can find new perspectives. For example, I held a group workshop for an international team of archeologists. We ended up on the moon, looking out on planet Earth. Participants reported a feeling of awe, that overview effect we brought up in Chapter 3. What a feeling, to share it firsthand together.

We can gather. I invited a business team to meet with me in a virtual world to learn about wellbeing before going on a virtual field trip as a group, in a teambuilding exercise.

We can immerse. Immersion can be more than a metaphor, as we travel to underwater worlds only possible to access in this way through shared spatial experiences. A community gathering turned into a dialogue that was an ideal way to prompt new ways of thinking and understanding.

We can go beyond the self and other, experiencing what it feels like to understand and be understood. Immersive worlds are powerful places to exercise emotional connection, empathy, wellbeing, collaboration, and resilience in a state of open wonder and awe.

We can give stories new meaning, using objects in new ways to help us to connect, to remember, and to bridge semantic gaps. We can also use these methods to understand data and information in new ways, mapping relationships, changing scales, looking at past and future, spanning the globe in an instant.

These are just a few examples among innumerable applications for immersive imagination, across all verticals. The key is curiosity. Curiosity and openness lead the way, and these are aspects of wonder that keep us adaptive and resilient on this spatially empowered journey.

I believe in the power of the metaverse to bring people together. It's the ideal place to stretch the imagination, to level up emotional intelligence, to deepen empathy and understanding for others, and to enhance collaboration, wellbeing, and resilience. In an article about the metaverse, I said,

> *The metaverse can give us better interactions, by design . . . for those of us ready to embrace and shape a future that allows us to connect in new ways that deepen dialogues and enrich understanding across perceived divides.*
>
> *The future can be bright, and the metaverse we dream about is only possible if we align it with our deep values and sense of ethics, and keep checking in with our humanity. This can be a starting point to a fuller, connected experience.*[1]

While we hear a lot about the shared, spatialized, and social aspects of the metaverse, sensory, soulful, storied, and somatic elements also play a key part in imaginative immersive interactions (Figure 8.1). We can apply this to better define the metaverse and deepen our considerations of how we can reshape the future of work and play.

Sensory

Metaverse **sensory** technology, in many ways, can be described as flipping your brain inside out and then inputting all of the context and sensory details that allow us to make sense of our situational place and form patterns and models. What about the senses that are missing? The textures? The scents? Oftentimes, our brains fill in the gaps for what's missing. We form a complete picture in our cognitive mental model, which affects our physical, emotional, and mental experience.

When auditory sounds are mapped in a certain way, they can create a texture of their own, eliciting emotions and responses that are different from the way we experience physical reality. In a similar way, visual cues and inputs, virtual settings and landscapes, become part of the fabric of how we encounter the metaverse, cueing our responses, our navigation, our agency.

Sensory elements can be things of beauty, inviting us into a state of reverie or wonder. When our senses are nourished, we are able to form better qualities of experiences. With wellbeing and social-emotional goals in mind, this has big impact.

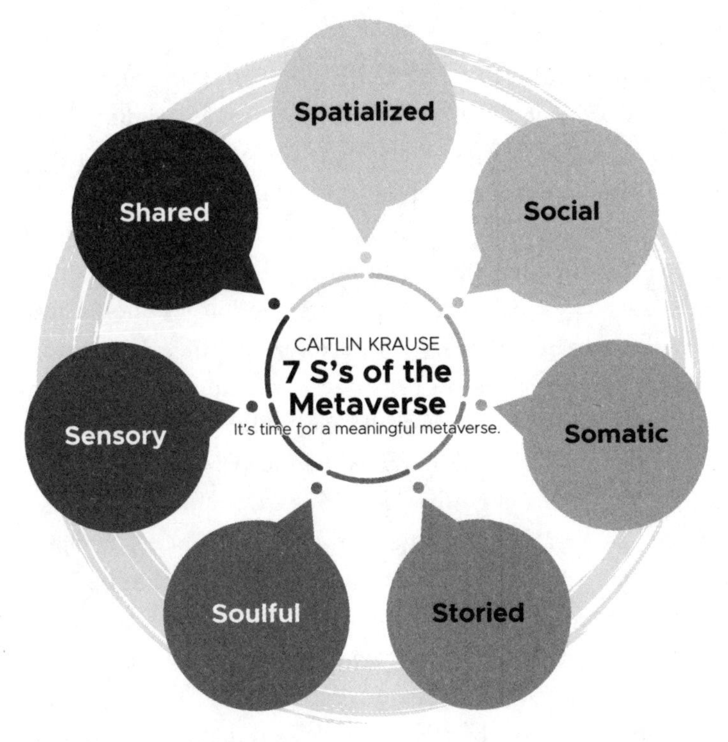

©Caitlin Krause, 2022

Figure 8.1 The Seven S's of the Metaverse.

Source: Caitlin Krause, "The Seven S's of the Metaverse," *Medium*, July 5, 2022, https://caitlinkrause.medium.com/the-seven-ss-of-the-metaverse-23a8bdd318b6.

Soulful

The word **soulful** is complex, and I don't mean to use it in a religious sense, though you could interpret it spiritually. What I mean is that the metaverse involves instantiations of the "self" in a collaborative context. So, who shows up there as you? You might be a certain avatar, you might be using new types of tools, you might be interacting across different languages, using a new form of what I call *metaverse fluency*.

As we encounter a metaverse that is so interactive and moving, we can benefit from awareness and guidance about how to show up as a soulful self, "full of feeling and full of emotion," as the definition states, who can make the most of the experience, for ourselves and for and with each other.

Storied

It's a **storied** metaverse that helps to make it meaningful. We as humans share stories with each other to transfer data, understanding, history, and more. We remember stories—hence the memory palaces that set the metaverse apart!—and we care about them. The metaverse is a "place"—a space with meaning. We can be the story makers and the meaning bearers. This is something that humans do. It's not generated by AI—though AI can help assist us as story creators, which has tremendous business impact.

Focusing on brand, stories we want to engage in, and how to sculpt a powerful user journey allows us to connect with audiences and communities. Humans need storied rituals in all facets of life, and 2D interactions just don't cut it. The metaverse allows for these deeper storied connections to take place in personal and professional contexts.

As we are facing a new way of integrated work-life design, rapidly shifting our approaches to business, to our daily routines, and to how we gather and celebrate together, we have the chance to approach stories in a new light too. Humans are social animals who exist as collectives, in community with each other. The metaverse is a space for sharing, offering us chances to focus on cooperative models and systems designs that enable us to thrive. By integrating stories in the metaverse, we raise consciousness about engagement design and agency that allows us to form fuller understandings.

Somatic

How does **somatics** play out in the metaverse? It might be instinctive to think that when we put on a headset, everything that happens in a metaverse happens in our heads, from the neck up. The immersive experience is not contained in the brain. The body is constantly responding to the experiences we have in the metaverse because it has opened up our sensory inputs, and valuable signals are coming to and from our physical bodies that can lead us to a fuller experience. Visual signals, auditory, haptics, and more. It's about the mental and physical together. Not exclusive.

Increasing awareness that the metaverse is somatic can allow us to understand why certain experiences give us goosebumps, or why breaking past limiting beliefs by giving a public speech in a metaverse world causes our heart to beat faster and perhaps our stomach to feel those butterflies. It's a full experience, and this is ironic because even as it's dubbed a simulation, our full body recognizes it as "real." There's a vulnerability here, tuning in more deeply to the body and the brain as they are connected. We allow ourselves to be physically moved by an experience. Oftentimes, the emotional responses in the body in metaverse will bring us to laughter, to tears, to a deeper understanding of ourselves and others, often across perceived boundaries. It's different than viewing a movie because there we are, in 3D, reacting and interacting. Imagine the powers we can have, as authentic leaders and learners, to work on transformations, when we also understand our own body integration.

Experiences in the metaverse, as we have seen, are more than "shared" and "social" and "spatial"—they are also sensory, soulful, storied, and somatic. In addition, they are built upon infrastructures that incorporate AI. Spatial computing and AI are intertwined, as the metaverse is powered by AI. It's an ideal convergence if predicated upon trustworthy frameworks. That is also why digital wellbeing is critical. Recognizing how these seven facets of the metaverse interplay and affect us raises our awareness about how to design better experiences with intention on our values and vision. We address the "why" of entering a metaverse in the first place, which is far more purposeful than merely replicating physical reality.

It's truly **connection** that is the point. Connection inside of ourselves, between us and other people, and that interconnection with nature. We are biological, after all. Layered, intention-driven digital systems can help give us the means to enable even more creative connection, if designed well, inspiring us to care for the physical world. Now is the time for that intentionality to drive the future of our immersive applications.

How XR Empowers Remote Connection

How many of us have to navigate hurdles involving connecting distributed teams? Andy Fidel, founder of Spatial Networks, knows the challenge well:

> *XR has made me value and strive for better remote experiences. It allows us to connect beyond physical limits in ways that are interactive. This improves my wellbeing by giving me more chances to connect and experience new things. . . . Now, with immersive digital technology, we have the opportunity and*

"We used to categorize events as On-Site or On-Line. Today, we have a third option—In-World."

> *tools to reimagine the online experience by reintroducing the concept of 'social presence' and 'shared experiences.' We used to categorize events as On-Site or On-Line. Today, we have a third option—In-World.*
>
> *With XR, location becomes secondary; you can be fully immersed in an "in-world experience" (in X). . . . Whether I'm in a bustling city in Bangkok or a remote village in Cholula, XR allows me to remain fully engaged in my work and connected to my colleagues and clients.*
>
> *I've always worked remotely with companies. This has made me appreciate teams that offer inclusive online experiences and opportunities. For me, it's about merging work and life, bridging the gap between the physical and virtual worlds. This exploration brings me a profound sense of fulfillment and wholeness.*

Fidel acknowledges that a full life of digital wellbeing includes defending mindful tech breaks: "While XR has brought many benefits to my work and life, it also has some drawbacks. More technology means more screen time, and this can sometimes pull me further away from a natural, unplugged state. To address this, I've had to stay aware of how much time I spend with screens. It's a balancing act."

XR hasn't distanced me from the physical world; it lets me curate physical experiences and make conscious choices. Routines can be different with the incorporation of XR. It impacts the work itself, enhancing freedom and agency, as we've seen from Fidel's approaches. It also makes it imperative to prioritize choice and to use care and attention in employing these approaches in our workplaces. The mention of boundaries is important. I've talked in this book about setting up intentional "speed bumps" as we're navigating seamlessly between spatial environments and physical ones.

We don't want to toggle task to task, bouncing around with no chances to take a breath and integrate learnings. Contemplating previous experiences, just as we anticipate the future, enables us to apply intention to what's about to come. On top of all of that, using deeper quality of presence to show up where we are with all of our senses lets us arrive and fully experience the present moment, which is essentially all we have in this life. That's a mindfulness tenet: to appreciate the brightness of "right now."

We don't want to spend our precious now moments always reminiscing or ruminating over the past or anticipating and hoping for the future. We want to dwell in the nourishment of the present moment. This is a hard one for humans. Can XR help? The accessibility the technology provides should not mean constant availability. Connectivity can mean we are over-inundated with expectations from our colleagues, but it doesn't have to be this way. The choice is ours.

AI and Spatial Remote Workflow

Choice is also broadened with the advent of AI integrations that we can apply directly to work. I asked Fidel about her daily routines, adaptation to new forms of work, and calmness in midst of AI-related bustle. "I boarded the AI train as soon as it hit the consumer market," she said. "Overall, it has made my days both busier and less busy. Essentially, it's given me back time to further invest in other expansions. So I wouldn't say I'm less busy, I'm equally driving forward, but at a faster rate. It's given me time to play and experiment."

I know from experience how busy it is to design and host live events, especially with content development. I prioritize my focus on participant experience. There is a lot of collaboration and coordination, and AI tools can increase efficiency and allow for creative focus and freedom on the part of the human designers, hosts, and creative leads. I project this will only get better as AI tools improve and more people understand how to integrate spatial technology into their workflow—and life flow.

I asked Fidel what advice she would give to others about using XR and AI to advance life and work. She said, "When using AI and XR, it's important to have clear goals in mind. Start by asking yourself what you want to achieve and determine which tools can help you reach those objectives. A world of infinite choices can easily become a burden and even anti-creative. So take a step back and connect with the intention in mind."

This resonates. In the past, I'd take on new roles for my business because I saw them as skills that I would benefit from learning. I thought that understanding all of the ins and outs of every aspect of having my own company made me better at my business. I still do have that overall mindset, but I have learned to delegate to AI and to team with other humans. I release myself from the urge to perfect everything myself, especially tasks that AI can support. I couple this with focusing on the collaborative work that can be more exponential and expansive with a human team. Four parts make a difference: opening up to imagination through wonder and awe; remembering to be both visible and a good listener; using emotional intelligence to empower meaningful human connection; aspirationally dreaming about larger objectives that we might not be able to accomplish as lifelong soloists. These tenets can help empower the future of human-centered AI-integrated work that supports our wellbeing through a lens of wonder.

I know it's challenging to achieve these things, along with supporting our creative and financial freedoms as entrepreneurs, but that doesn't mean that this level of empowerment is not worth striving for. It's vital and necessary, and the time to have deeper conversations about the "how" is right now.

Because spatial XR is speaking a universal language of 3D object-oriented experiences, it's accessible to a lot of different groups across perceived divisions. It can bring people together and recontextualize what it means to be human. When I asked Fidel what her world is asking for, she said, "My world is asking for deeper connection. It's not about speed, visual quality, or format, whether in the form of a letter or a hologram. It's about sharing meaningful experiences and having more authentic and genuine relationships."

Spatial technology offers opportunities to explore in experiential ways, solo or socially, in immersive spaces that offer a chance to *feel* new ways of knowing, which is much more powerful than thinking. As Fidel says, "Spatial computing and XR encourage diverse perspectives, create safe spaces for exploration, and empower active participation, fostering growth and experimentation. They have the potential to enhance and expand our reality."

Flow States, Human Imagination, and Spatial Possibility

I've often said that "spatial is special" because it gives us a chance to experience wonder and awe, closely linked to the flow state experience, in 360 surround scape. It's beyond our natural habitat as humans—it's our supernatural habitat, and we get to exist in superhuman form. Let's look at the concept of flow, recognizing convergences.

In 2019, I gave an experiential talk about VR called "VR Is Visceral" at SXSW EDU with Steve Dembo. During the talk, we showed "The Presence Pyramid" of learning and experiences (Figure 8.2), asking participants to describe, in stages, how they learn and feel experiences in meaningful ways. Educators know this because it's part of our training: we can start by

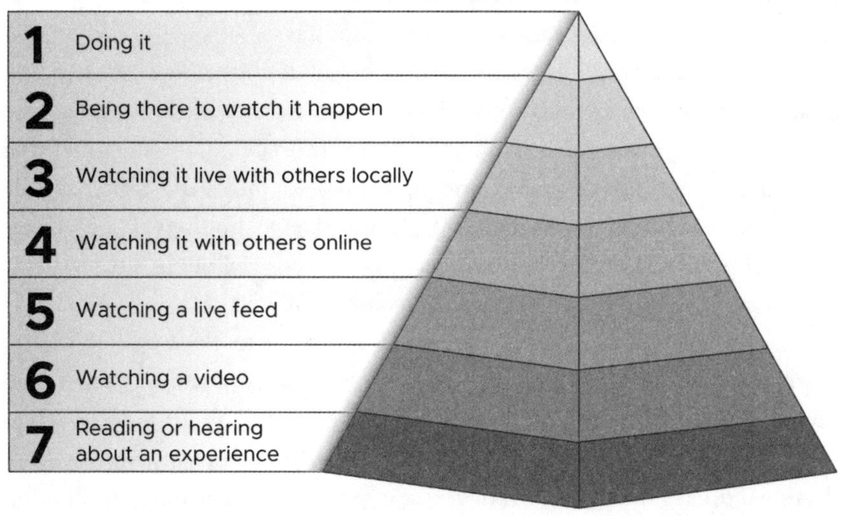

Figure 8.2 The Presence Pyramid of engagement and experience design.

reading about something, then we can learn by talking about it with someone else, and we can visually see it acted out for us. As we have been describing and referring back to throughout this book, the most meaningful and powerful transfer is *experience* itself, when we *do something* and have a full-body-awareness of the experience. This embodied doing in copresence with others can be even more meaningful because we belong collaboratively and co-creatively *in* the process.

When you can also have emotional engagement as part of this experience design, it becomes even more salient and meaningful to you. The interconnectedness of your place and belonging within the experience encodes it in the brain and body, which will serve as experiences from which you draw and reference in future real-life scenarios. Now, in this "bead on the necklace" of seeing how much spatial experience can change life thriving and overall wellbeing, the full power of possibility becomes evident.

In terms of healing, those looking to overcome a phobia or addiction, dealing with personal setbacks, or seeking to reshape stories, spatial experiences can have powerful therapeutic effects for the entire body and mind system. They can also be the place you go to in order to reengage your senses, refresh, and spark new innovative mindsets.

These understandings led me to deepen my focus on wonder, awe, and imagination, which incorporated research and work with flow. A concept popularized by researcher Mihaly Csikszentmihalyi, flow is something I've written extensively about and incorporated into XR design. While Csikszentmihalyi never worked with immersive technology, I see XR as the ideal place to develop these fluencies, blending wellbeing, imagination, and collaboration skills professional development.

Instead of having a rigid, systematic approach to flow, I offer frameworks and methods that are more curiosity-motivated, open, expansive, and interactive. I feel it's important to avoid being clinical about flow. Much as we might want to "optimize our life" and apply flow optimizations to everything we do, it's important to invite these powerful experiences to develop integrally and to operate from a centered place of letting that openness invite what is emergent in ourselves. That is flow.

For example, you'll notice in Figure 8.3 my mapping of the connections between traits of flow. Many of these traits were identified by

Figure 8.3 Connections I see between awe, wonder, immersion, and flow.

Csikszentmihalyi—and, by incorporating awe, wonder, and immersive worlds, the traits become perfect complements for what happens in a spatial experience. Spatial is the special place where you can practice flow and enhance your "flow muscles." It does not have to be sparked by an extreme sports experience. This channeling of flow can happen through many vehicles that immersion is especially good at: visuals, auditory cues, haptics, and more. Spatial technology can be designed as a "flow priming" in a sense. In XR, across a spectrum from AR to VR, I have built spaces and experiences that I call "Flow Camps" for teams and individuals alike. What does it feel like to experience awe and wonder in an immersive world, transferring flow back to your physical reality? It's like training body awareness, for a sport, and that embodiment can have a lasting effect on quality of life in all areas: personal, business, academic, industry, health . . . everything is in connection through your moment-by-moment experience of flow.

Taking Flying Leaps to Reach New Heights

James Jensen, creator of JUMP, has a vision for flow, aiming to help individuals overcome fears and connect deeply with virtual experiences.[2] We had a conversation about awe, flow states, and breaking through perceived limits. Jensen shared, "We can give the experience of skydiving and wingsuit flying to a 10-year-old. To do this in reality would take years of risking their life and thousands of dollars, and they would only be able to begin once they're 18. At JUMP, they can step up to that decision, embrace all their fear of failing, jump past that doubt, fly in celebration, and experience success. It can be a life-changing moment for a child struggling with confidence, stress, anxiety, self-worth, depression, or any other emotional disability that could be rewired with experience."

It's a brilliant point that we've made before: when it comes to the possibilities of XR and spatial technology, we're rewiring the future. Jensen has introduced a way to encourage people to overcome fear and experience flight. What does he think is our greatest asset as we strive to rewire some of our own patterns and open up to new ways of being? Jensen says, "Overcoming these self-imposed personal limitations would free a person to strive for their full potential in whatever they want to achieve." From overcoming phobias to setting new goals and visions, spatial technology makes so much more possible.

The End Is the Beginning

This is the beginning of a new era of spatial exploration in the name of **freedom and connection.** There is much to anticipate and value. As you take time to reflect and respond, let's also consider the possibility of expanding our conversations to immersive spaces to have forums and interactive discussions about these topics as we are experiencing wonder.

Keeping in mind what Rodney Mullen has shared, "The process will involve unlearning and openness as we explore dynamics that are more fluid than static, more integrated than separate."

This points to the essence of flow.

Personal Reflections

1. How can digital spaces feel as fluid as physical ones, without needing to fully replicate the physical? What elements of physical experience bring you into the greatest state of wonder, connection, and flow with your environment?
2. How do we address the paradox that more connectivity is leading to increased loneliness? What better XR tools and ecosystems need to be built to respond to this challenge?
3. Why does it seem as if humans are becoming more robotic in the ways that they approach life, while AI systems are increasingly becoming more human-like? What does this say about the future, and what is important to keep in mind?
4. What boundaries would you like to break, and what personal mindset shifts would you look to empower, through the use of imaginative immersive tech?
5. What are your reflections about the seven S's of a mindful metaverse? How do you imagine it would feel to be part of a metaverse experience that prioritizes all seven?

Reflections for Leaders

1. What have your work-related learning experiences been like when it comes to the Presence Pyramid? How do you learn best and form meaningful experiences? What would it look like to implement aspects of the Presence Pyramid in your business?

2. How can we use spatial technology to prepare ourselves and future generations for long-term resilience in this rapidly changing landscape, where new jobs and skillsets are needed, most of which haven't yet been defined?
3. We've heard in this chapter that remote work is easier because of spatial technology. How would it help your leadership and empower distributed teams to incorporate aspects of spatial technology?
4. What do you think of the statement: "Let's focus on rules of engagement before tools for engagement"? What collective norms do you have in place for the way you and your teams engage with each other? How do you foster better engagement strategies?
5. How could empowering better uses of spatial imaginative immersive models transform everything from customer experience, retention, work satisfaction, and beyond? Here, the sky—the galaxy—is the limit. What would you like to see most impacted when it comes to your future of work and wellbeing?

Extensions for Further Exploration

There are many ways to expand on this chapter. I hold up the power of immersive imagination and would like to keep extending the research and focus of this chapter on wonder and spatial technology by continuing to develop shared learning opportunities, in business and education, both personally and professionally. Research will be expanding in the coming months, and more organizations will be adapting and transitioning to AI-enabled and supported XR environments as part of the future of work. They will need creative, imaginative solutions as well as meaningful support in these transitions. Not only do they need hardware, they need experience onboarding and content literacies as well as better workflow and team collaboration understanding in these new systems and environments. It's going to be a tidal shift! Some places see that "next" phase coming, yet they still think in 2D ways about technology. Expanding here to more experiential 3D discussions bridging those topics and offering ways to rise to a higher plane, together, is part of where I would like to lead the conversation—at global scale. There is great possibility, and it's time to build a movement.

9 Wellbeing Through Music: The Heart Is a Digital Kick Drum

Think of the last piece of music you listened to. Was it on Spotify? Or maybe on your record player, broadcast through a digital speaker? Was it in your car? Could the composition have been created by AI? The bead on a necklace in this chapter has to do with (drumroll) the most powerful human sense for wellbeing, which is now distributed, aka, mediated, broadly in digital form. Music can help us in so many ways, and digital wellbeing through music and audio cues has limitless possibility. Let us explore how and use examples that can apply to our every day.

In the United States, a staggering 85% of audio consumption comes from music streaming services, while only 15% is from owned music.[1] This clearly indicates that the vast majority of music consumption in the United States is through digital means. Additionally, US CD album sales reached 40.6 million units in 2020, which was a 26.3% decrease from 2019, further highlighting the decline of physical formats in favor of digital ones.[2]

Globally, the music streaming market size was valued at $24.4 billion in 2020. Worldwide users of on-demand music streaming services increased from 57 million users in 2014 to 419 million in 2020,[3] a testament to the global shift toward digital music consumption.

With digital wellbeing in focus, it's a wonderful time of possibility, as music is able to be recorded and transmitted through devices, reorganized and framed digitally, sped up or slowed down, and transformed in new ways. AI applications are also transforming music creation possibilities, and our abilities to play both creator and consumer of music is more powerful than ever. With all that in mind, let us investigate the power of music to enhance wellbeing, approaching ways to integrate it into our digitally mediated lives.

The World Is Sound

In 2018, I visited the Rubin Museum in New York City, one of my favorite places, and I sat down for coffee with deputy executive director Tim McHenry. We talked about sound, the museum's exhibit at that time.[4] They had hosted a museum sleepover the week before—can you imagine the wonder of being in the Rubin overnight, surrounded by Himalayan artifacts, statues, paintings, and tapestries? Pure magic.

During the sleepover, guests wore blindfolds for a sound isolation exercise, paying attention to their somatic body experience even better. When humans first hear a sound, our instinct is to try to identify it by connecting it to a memory.

What McHenry and I talked about at the Rubin was the art of paying attention to the *quality* of sound itself, without assignment of a label. What is it like to simply listen to the pure sound of sound itself?

That conversation has stayed with me ever since and made me think in a new way about sound. I started to focus on the mindfulness exercise of appreciation, of looking at sound as layers of emotional resonance.

Poetry is one of those remarkable multilayered creations that allows humans to connect and share in ways that can combine with computing technology and reflect music. I have been giving "AI and creativity" expressive workshops for a decade, calling them: "Humans Beyond Robots: The Heart Is a Digital Kick Drum." The sessions are all about integrations with AI, biorhythms, emotional intelligence, and digital wellbeing. These experiences are therapeutic and creatively expansive. I want humans to access imagination and see what Marvin Minsky once pointed out, that it's all a giant ecosystem. Our interactions with digital media, music, and AI are part of the ecosystem of being human and having access to beats and emotions that guide us. It's an inside out provocation of possibility.

In 2023, at the annual mainstage TED conference in Vancouver, I was asked to lead a discovery session about XR and emotional intelligence. Involving poetry, story, and music in integral ways was part of my session design. Participants came from a range of backgrounds, cultures, and disciplines, and primarily held leadership roles in their businesses. We focused on music and poetry expressing our humanity, the language of emotion, and the collaborative power of tech-driven spatial storytelling that has a foundation in wellbeing and connects with the physical body. It was a deeply imaginative session.

We are creatures of music, of internal drumbeats. We each have a heart that beats in an iambic rhythm. Stress, unstress, stress, unstress. . . . Music, sound, and neuroscience are interconnected in ways that significantly impact wellbeing. Research has shown that music can have a profound effect on mental health, the brain's processing of sound and music, and can be incorporated into a wellness routine in various ways.

Music is often considered the ideal sensory experience for digital wellbeing due to its ability to evoke emotions, regulate mood, and foster social connections, among other benefits. Let us explore more of the "how" in this bead on the necklace.

The impact music can have on our wellbeing is absolutely astonishing. Here are just a few ways digital can be beneficial to our health (this is true for nondigital music, too, yet digital streaming services and platforms make digital so accessible that it's essentially a mainly digitally mediated experience these days): It can act as a distraction, reducing stress and lessening the impact of depression and anxiety. Listening to music can also boost mood, with research suggesting that listening to upbeat music can serve to shift disposition.[5] Music therapy has shown promise in providing a safe and supportive environment for healing trauma, improving mental wellness, reducing stress, and alleviating pain. It can also facilitate emotional release, promote self-reflection, and create a sense of community.[6] Music as therapy can also be used for physical wellbeing. It can help organize and structure movement to improve efficiency, as in dance choreography, and help us learn and relearn new movements quickly.

Incredible, isn't it? We *feel* this as true. Just think of how different watching a film is when you turn the sound off. Not as moving, right? Music is a way of understanding and accessing meaning, beyond words. In terms of music and the brain, listening to music activates the entire brain and can be used to shift both the brain states and the states of the body. It's not isolated as an auditory processing experience. It's an emotional experience.

When we evaluate music and movement combined, we find that music can prompt motor actions that occur throughout the body. Arms, legs, torso, head, *everywhere* in the body, music, movement, and dance are intermixed. Knowing this, we can better tune for entrainment, which we discussed in earlier chapters as part of the Seven ThEmes of spatial computing.

Music prompts movement, and so we are entraining for engagement as well as for emotion, by design. We can tune the whole body to be even more receptive to its own wellbeing, its natural flow state.

Music for Motivation

Music can motivate when we want to get something done. Everything we do has a certain rhythm to it, and the pace of music (broken into "beats per minute," or bpm) can cue us to follow that beat and motivate us to move. One tip is to listen to music between 140 and 150 bpm if we want motivation to exercise or perform cognitive work. I use certain wordless music, between 50 and 120 bpm, to lead me into a "flow state" of creativity.

Listening to music for productivity needs to be very precise because it can become distracting if not chosen with care. I've been practicing using music to augment my workflow, taking movement stretch breaks from my flow state writing work that I do with wordless music. The breaks help to reinforce Parkinson's law, which states that work expands to fill the time available for its completion.

To combine interval focus techniques with music, you can use playlists that are specifically designed to complement the timing of regular sequential breaks. For example, designing 25–50 minutes of energizing focus music followed by 10 minutes of relaxing break music can work well. You can create a playlist that fits your preferences as a digital wellbeing exercise.

Pairing intentional music with specific activities can be very effective in regulating mood, too. It takes time and care to curate and then refine to maximize the impact, utilizing bpm as a guide. Knowing that there is science underneath that grounds the process is part of what I find so fascinating.

I use deliberate breaks in my flow techniques, along with music, to tune my body for the work it's doing in digital worlds. The small breaks serve to allow me to recharge and keep myself accountable for the larger goals, and they also prompt me to disengage from social media, which can be distracting.

I use music and movement breaks when I work in stillness for long periods. As leaders, we can encourage our whole team to adopt these breaks too, which support better engagement and productivity. Cal Newport, author of *Deep Work*, recommends specific methods including structured breaks to support sustained focus. He acknowledges that while deep work requires intense focus on demanding tasks, few people can sustain peak cognitive intensity for more than an hour without some form of relief.

Our digital infused lifestyle, which can often command our attention on screens and across a vast continuum of immersive devices, demands that we adopt new rituals to maintain our best quality of relationships, imagination, and health.

When it comes to music and productivity, Newport mentions that people can habituate themselves to different types of music, which then becomes part of their work ritual, indicating that the ritual of listening to music can help trigger a focused work state.[7]

I agree. For certain writing tasks, this chapter included, I have wordless playlists that are curated within a certain bpm range. It's much like entering a state of meditation. I can tune in and tune out at the same time because I've habituated myself to the music.

As we investigate what methods work best for focus, I should mention that when it comes to learning and cognition, some may find that *silence* helps focus and flow. That said, surfing wonder and creative imagination can also fall into a different realm altogether, and form fits function when it comes to music and wellbeing. In other words, find what feels good. In all of these cases, there are also many people for whom silence can be uncomfortable, including those who have experienced trauma. These ideas and exercises are meant to be supportive, so please check in with what works best for you.

Digital Wellbeing Through Music and Poetry

Music is a language beyond words. I would call it its own fluency. In this digital age, it is constantly mediated by digital technology. Music is media.

Music, along with singing and dance, likely evolved before spoken language. The original stories, all in spoken word, were passed down through poetry, which is spoken music. After producing a range of poetry and music across digital media, I do not think it is as linear as "happy music makes us happy; sad music makes us sad." There are more nuanced interpretations.

I have been wondering about music and emotion ever since hearing Elton John sing "sad songs they say so much." For me, listening to a "sad" song can uplift and inspire me. It reminds me of my place in the world and of universal suffering. The music moves through me and allows a shift to take place, an emotional reawakening and remembering of a truth. I move from that place of truth. That is one of the best ways I can explain it.

There are no positive and negative emotions, only a wide swath of variations and complexity, and everything is to be experienced with openness and compassion. Music can help foster that self-awareness and better understanding of others and the world around us. Music is moving for a reason.

Music expands our emotional capacity because we inhabit it as it inhabits us. It's a total mood shift that opens our minds and our hearts. In this musical digital wellbeing bead on the necklace, let us find out more about how and why it can have such a powerful nearly instantaneous effect on our wellbeing.

Expanding the Mind Through Music

The brain and body respond to surprises and changes in patterns. We are predictive creatures, always on the lookout for patterns, which then cue our actions and our emotional labels for experiences. We can shift all of this through new music, which has a way of surprising and delighting us. We can use knowledge about music to "shake it up," introduce new tunes, and support the brain's neural circuitry and neuroplasticity for learning. In fact, many musicians are also talented mathematicians. It's not about divisions in left brain versus right brain; it's about enhanced capacity in neuroplasticity that music trains the brain to support.

What we hear and feel in music can imply intent, which is why it is used to such powerful effect in media, from digital games to films to commercials. The soundtrack is our emotional tenor and cue about what we are supposed to be feeling. We use these cues in the ways that we interpret music. All of it can be better understood and applied to wellbeing applications.

When our neural circuits are activated by music, we feel a call and need to respond through movement of some sort. Have you ever been at a concert and found it impossible to sit still? Also, the area of our brain responsible for speech neighbors the motor area for our hands, which influences why we "talk with our hands"! The integration of these networks is at a biological level. It's incredible to discover about all of the ways we are designed and coded to create harmony, synchronicity, and belonging with the world around us. We are instruments.

Expanding the Heart and Body Through Music

Think of all the classic movies where music and dance were forbidden. It's the premise of *The Sound of Music*, *Footloose*, *Dirty Dancing*, and countless other classics. These were joyless, dispassionate places before music came on the scene, right?

Science backs it up. We have an overall positive physiological change when we listen to music even just 10–30 minutes each day. As mentioned, reductions in resting heart rate and increases in heart rate variability endure for hours after listening to music. They activate features of the parasympathetic nervous system that are lasting.

Our heart, and all of our bodily processes in fact, are naturally tied to our breathing patterns, so it's also postulated that much of the physiological benefits and heart-enhancing experiences are closely linked to our breathing patterns undergoing subconscious changes when listening to music. Music, wellbeing, and breathing are all linked.

Music Daily Practices for Wellbeing

Music and sound can lead to various health benefits. Listening to music with a slow tempo, for example, can decrease blood pressure, heart rate, and respiration rate.[8] This can help put the body into relaxation states, aiding natural healing.[9]

Specific techniques can enhance vagus nerve stimulation while listening to music: focusing on deep, rhythmic breathing and mindfulness, combining listening sessions with meditation or yoga, and using headphones for a more immersive and focused experience.[10]

When it comes to wellbeing and connection, certain cadences can lead to flow states. It's up to us to experiment with what works best. I prefer to start the day in silence, with a short meditative practice. Silence is never absolutely silent, and every sound, even our own breath and heartbeat, becomes music.

The connections between digital wellbeing, music, poetry, and sound are powerful in their possibility to enhance daily life thriving. The neuroscience of how the brain works shows that music and poetry can be tools for promoting mental and emotional health, and understanding how the brain processes music and sound can provide insights into its therapeutic potential.

Personal Reflections

1. If you had a playlist for your life, what would be included on it? What would be your "go-to" songs for different mood states?
2. How have you experienced poetry in the past? How would you potentially like to incorporate it into your life, now that you know more about its wellbeing properties?
3. Which techniques of digital music therapy are most intriguing for you?
4. Have you ever tried using music to motivate you to do anything, from sports to a cognitive task? What were the results?
5. Did you play a musical instrument growing up? If yes, what was the experience like for you? If no, what would you have liked to learn to play?
6. Why do you think we call it "play music" and not "work music"?

Reflections for Leaders

1. Music and work have often been two separate topics. How do you see the two relating now?
2. How can music be used for entrainment at work, to invite you and your team into a better creative flow state? How would this affect the cadence and pace of the day, with Cal Newport's findings in mind?
3. If you could create a playlist that reflects your leadership style, what would be on it?
4. Knowing that vocal cadence and musicality is part of what can be the greatest influence, how has this affected your public speaking skills as a leader?

5. Have you considered ways that you could use music to increase wellbeing or productivity during the workday? What are some of your ideas and intentions?

Extensions for Future Exploration

I would like to further research how digital music is interacting with sound practices for healing and imagination.

In addition, poetry is a tremendous area for expansion and exploration—as a creative practice, as a vehicle for wellbeing, and as a beautiful experience for listeners because it blends with many art applications.

Bringing these experiences into dome settings is another arena I have been developing. For those interested, you can follow more of my ongoing projects at caitlinkrause.com.

This whole chapter becomes expanded through possibility. Let it go; let it flow.

10 Wellbeing Through Wearables and Movement

We were built to be kinetic, meant to move and interact with the world around us. This is part of why we have stereoscopic eyesight, among countless other physical attributes, including our biped mobility and our level of manual dexterity. We were meant to both inhabit and manipulate the world. This bead on the necklace addresses movement and digital wearables that support our overall wellbeing.

As you read this chapter, I encourage you to look at your own bodily experience in the world. I view my own body with a great sense of dignity and gratitude. It's the way that I experience everything around me, taking in signals through my senses and processing them to form experiences and make meaning out of those experiences. We are biological creatures wired for wonder.

Our biology is the base of our wellbeing, and wearables have a potential to be layered on top of our biological frameworks in ways that can serve two prime purposes: they can better inform us about our patterns, habits, biofeedback indicators and levels, and they can also serve as interventions to help us manage our behaviors. In this second category are two further subcategories: those devices that will "nudge" us, giving us subtle cues and help with redirecting our behavior, and devices and wearables that will provide treatments and augmentations, serving as digital supplements, in a sense.

At the outset here, it's important to name that when we talk about wearables, we are essentially talking about the use of sensors to gather data. We talk about them here alongside algorithms, which extract parameters from the data like heart rate, temperature, etc. We also form longitudinal insights, where we see what factors to change in our behavior. We use data from the wearables over time as a measure for how we are progressing. This is why it's so valuable, at this stage, to consider the user interfaces, the wearable form factor, and also the reliability of the algorithms. All in all, wearables have a potential to enhance wellbeing, in my opinion, if we do not get too caught up in the numbers or fixated on the data. A paradox is at play here, where we want to make sure our attention is not on the devices as opposed to the experiences.

When it comes to wearables and wellness, there's not one device or one way of tracking or monitoring that should be a "go-to." I believe in the tenet "form fits function," and our lifestyle and goals should help us to make quality, informed choices about whether a wearable is helpful to serve our overall wellbeing. Just as wellbeing itself is a dynamic construct based on context, wearables are much the same in how we can approach them.

Just as wellbeing itself is a dynamic construct based on context, wearables are much the same in how we can approach them.

This book does not attempt to set quality standards for specific devices and products. What I will mention here are broad categories and examples of use cases that are well documented. In the future, I may attempt to provide reviews and add my personal opinions about what seems to work well. In no way do I wish to endorse specific products here, as that is explicitly not in the scope of this book. I do, however, think it's useful to understand and share about the categories and language that these products are using. They are disrupting the landscape of wellbeing, as digital wearables are being integrated into many people's lifestyle choices.

More people are choosing to supplement their life with a range of wearables that serve to give them freedom of choice, creating awareness about interventions that are possible and augmentations that could help their wellbeing. These wearables also provide a sense of agency, as people can choose to support lifestyle changes through the help of these assistive devices and protocols.

By contrast, this personalization of digital wearables, and the overall "biohacking movement," have several associated considerations that could be drawbacks. First, they increase a propensity toward self-consciousness and possibly narcissism. Second, the tracking of statistics and features of biological health can become a fixation and possible addiction, with the possibility of contributing to over-exercise and eating disorders, among other extreme behaviors associated with self-harm. What starts as simple monitoring can become obsessive. Proceed with care.

Through the course of my exploration of wearables, considering their design and use cases, I also became aware of what is called the Hawthorne effect.

The Hawthorne effect is a phenomenon in which individuals modify their behavior or performance due to the awareness of being observed. It is credible and has been observed in various settings, including research studies and workplace environments. For example, have you ever changed your behavior when you knew you were being watched? Were you ever on a web meeting, for example, and felt a change in the energy of the call when suddenly, the meeting shifted to a recorded one? In the context of neurofeedback and biofeedback, the Hawthorne effect may influence the results of the integrations. For instance, individuals undergoing neurofeedback or biofeedback therapy may exhibit changes in their physiological functions or brainwave patterns simply because they are aware of being monitored rather than solely due to the specific effects of the treatment.[1]

Introducing Measures of Wearable Devices

Wearable devices are increasingly being used to track various human biological measures relevant to wellbeing, which can be categorized into three major areas: physical health, mental health, and sleep.

In the area of physical health, wearable devices can monitor a range of physiological parameters, such as heart rate, respiratory rate, body temperature, physical activity, blood pressure, and blood oxygen saturation. These measures provide valuable insights into cardiovascular health, lung health, and overall fitness levels.

Mental health is a complex category, as it is often tracked as a self-reported qualitative measure. Wearables can indicate mental health parameters including mood, focus, and levels of calm based on user inputs.

The parameters specifically measured by wearables for mental health include heart rate variability (HRV), skin conductance, breathing patterns, cortisol levels, and voice recognition. Fluctuations in the time intervals between heartbeats, as measured by HRV, can indicate stress levels and emotional states.[2] Skin conductance can be an indicator of stress or arousal. Monitoring breathing patterns can provide insights into an individual's state of calm or stress. Some devices in development also aim to measure cortisol levels in sweat, which can be an indicator of stress. Additionally, voice recognition technology is being used to track mood states and neurodegenerative symptoms as well as other state indicators.

Sleep is the third crucial aspect of wellbeing that can be monitored using wearable devices. Certain wearables can accurately track sleep patterns and quality, providing insights that can help users better manage their sleep.[3]

While wearable devices provide valuable data, it's essential to use them in conjunction with validated wellness assessment tools to provide a comprehensive understanding of wellbeing. It's also crucial to ensure the accuracy and reliability of the data, which depend on the quality of the wearable device and its correct usage. Furthermore, the interpretation and application of data should ideally be guided by healthcare professionals to ensure personalized and effective health management.

Demystifying HRV

One of my personal favorite measures of health using a wearable is heart rate variability (HRV). I've been tracking my own HRV, on and off, since 2018. I'm a long-distance athlete with a passion for running, and I first learned about HRV when I was interested in heart health and also understanding more about the importance of my own body's ability to rest. I'm the type of person who feels incomplete if I have a day without sweat, and I grew up as an athlete, playing sports in all seasons. It's hard for me to take a day off.

HRV is a measure of the variation in time intervals between consecutive heartbeats. This physiological phenomenon reflects the body's ability to adapt to various internal and external stimuli, including stress, physical activity, and rest. HRV is controlled by the autonomic nervous system (ANS), which consists of two branches: the sympathetic (activating) and the parasympathetic (deactivating) nervous systems. The balance between these two branches determines the body's stress response, recovery state, and overall wellbeing.[4]

A higher HRV indicates a more adaptable and resilient system, and lower HRV suggests a less adaptable system, potentially signaling health issues or reduced resilience to stress. High HRV is generally associated with good cardiovascular health, lower stress levels, and better overall wellbeing. Ranges can vary significantly based on individual factors, and in general, below 20 milliseconds is considered low and above 40 milliseconds is high, though adults can record HRV above 200 milliseconds. HRV can also be used to monitor the body's response to stress and its recovery, making it a valuable tool.[5]

Chest straps that use electrocardiogram (ECG) electrodes are generally considered the most accurate devices for HRV monitoring. A ring can be a less invasive option for monitoring HRV, and while smartwatches and general fitness trackers can measure HRV, they may not be as accurate as chest straps, particularly during physical activity.[6]

The Big Picture: Why Use Wearables for Wellbeing?

We've talked about a wide variety of wearables. In sum, these are all electronic technologies that are worn on the body, either as accessories or as part of our clothing. Let us continue to keep in mind that when we talk about wearables we are essentially talking about using sensors to gather data from our bodies and brains. We have many choices right now when it comes to wearables, and more will be developed in the near future. It's a rising market. We can consider the strength and accuracy of the algorithms, which extract parameters from the data like heart rate, temperature, etc. We can also think about how we form longitudinal insights, where we see over time what factors to change in our behavior. We use the data from the wearables over time as a measure for how we are progressing. The more comfortable, efficient, safe, and intuitive to use the better.

Many wearables are designed specifically for medical and mental health applications. Wearable ECG monitors, blood pressure trackers, and other devices can transmit patient data to healthcare providers. Mental health wearables monitor brainwaves and physiological indicators to provide insights into stress and emotional states. Wearables are also being used in workplace wellness programs to help employees improve their health and reduce absenteeism.[7]

Many of these types of devices are called "smart" because of their intelligence, yet it's the humans choosing to wear them who need to make healthy decisions based on what they indicate. Wearable technology is continuously evolving, with new devices and functionalities being developed to cater to a wide range of personal, health, and business needs. The integration of sensors, processors, and connectivity options allows these devices to collect and analyze data, offering insights and convenience to users in their daily lives.[8]

Wearable devices have unlocked a new frontier in personal health monitoring, empowering individuals to take a more active role in managing their physical, mental, and sleep wellbeing, among other areas. However, these technological tools are only as effective as the user's ability to interpret the data and incorporate it into a holistic approach to wellness. As we explore the role of movement and digital being as part of the next bead on the necklace, let us keep in mind that wearables will undoubtedly continue to improve, offering increasingly sophisticated insights to guide our path toward optimal health and vitality.

While we have looked at music and wearables for wellbeing, now we are looking more deeply at what our physical bodies naturally do well—move! We were designed for movement. Let your physical experience of movement help you as we address wellbeing and physicality in this bead on the necklace. It's a digital wellbeing essential topic because, as you'll see, digital media can integrate as vehicles for movement and could also become obstructions of movement if they influence a more sedentary lifestyle. Let us see how this all plays out in our own experiences.

Movement and Digital Technology

As we have emphasized throughout this book, our body was meant to move. Its design and form is in 3D, and we think and perform according to 3D concepts in a 3D world. Our imaginations are embodied and whole, and when we hear sound, it reverberates in us and we begin to inhabit that music. We are creatures who belong in our world as a part of it. Sometimes, discussions about technology become so abstract and separated from the context of our physicality that we forget that essentially our life, our wellbeing, is all about the experience of being in the world.

Sometimes . . . we forget that essentially our life, our wellbeing, is all about the experience of being in the world.

Let us start—after a necessary stretch break—with a big "why": Why movement in the first place? Why were our bodies designed to move, and how is digital helping or hurting us?

I began asking questions like this a few decades ago, when I would suffer intense neck pain while working at my laptop. I took breaks to go running or swimming. Exercise led to better energy, which led to better quality of work. Whenever I could, I timed my workouts right before a big academic surge. I ran to clear my mind, to think over concepts, and to reach a certain meditative state.

Reflecting on this, I began to wonder why my body wanted to move in the first place. Was it to experience freedom? I designed my classrooms around the concept of freedom of movement and flow: space for reflection and introspection; space for sharing; opportunities to get up and move rather than sit hunched at desks. This is part of why working in VR is so appealing to me: it's all about movement!

Recently, I listened to the Body Electric podcast and participated in the accompanying Columbia University study. The study aimed to investigate the impact of incorporating regular movement breaks into daily routines. More than 20,000 people joined the study. The research, led by exercise physiologist Keith Diaz, found that taking short movement breaks, such as every half hour, could counteract the harmful effects of prolonged sitting.[9]

The fascinating podcast addresses the growing concern over the impact of sedentary behavior and increased screen time on physical health. This "way of being" does not have to be the way! The study's findings highlighted the significance of habitual movement in mitigating the adverse effects of prolonged sitting, with implications for workplace norms, societal habits, and long-term health outcomes.[10]

My most important reflections after participating in the study are a set of my own five "body electric" truths and principles for movement and wellbeing:

1. The first principle involves what I would call an **overall personal reframing when it comes to movement and wellbeing**. Embracing tiny exercise habits in daily life, as advocated by B.J. Fogg, can significantly impact overall wellbeing, particularly when integrating movement into every hour of the day. Whether it's a walk, a dance, or folding laundry, these activities break the sedentary cycle and contribute to a healthier lifestyle.
2. Recognizing the **importance of rest and recovery**, which includes varied movement, is also crucial. High-intensity workouts are valuable, but they should not be an excuse for inactivity the rest of the day. Acknowledging the pressure to exercise and reframing it within the context of fit and flow allows for the development of sustainable lifestyle habits.
3. **Positive self-image and self-talk** are powerful influencers of wellbeing, and the digital age brings both opportunities and challenges in this regard. Reflecting on personal physicality and self-reverence is essential, as is the way we communicate with ourselves, akin to how we would treat a best friend.
4. **Breathwork** is a fundamental aspect of physical activity that transcends into mindfulness and meditation practices. As an athlete, I learned to harness breath during various sports competitions, later translating my knowledge into a deeper appreciation of the body's capabilities. This understanding reinforces the notion that we are inherently designed to move, breathe, and engage with our environment in a dynamic way.
5. The concept of "**singing the body electric**," inspired by Walt Whitman's poetry, encapsulates the essence of being at one with the body and soul. This symbolic learning underscores the interconnectedness of our physical and spiritual existence and the continuous evolution of our being. It's a reminder that we are all athletes in life, constantly moving, breathing, and shaping our existence through our actions and interactions with the world around us.

All of these messages are core digital wellbeing movement principles of mine that I was reminded of by the Body Electric podcast and study—and they leave ripples in the lake.

On Being and Moving

Why are we here, on the brink of the mid-twenty-first century, talking about digital wellbeing? What are we doing with our electric bodies? How is today a time that *Frankenstein* predicted, in the reanimation and question of what it means to be human? Our fresh retelling is one in which AI is the creature, isn't it? Who animates the creation, and who is responsible for naming it, training it, and treating it with belonging in ways the monster was not? Of course, I'm prone to draw parallels here to AI, with its electric allure in our modern sci-fi. What would Whitman say? Mary Shelley herself was inspired by the role that electricity plays in life. She drew from science. The concept of the "body electric" encompasses the fundamental role that electricity plays in the human body's functioning. This idea is not only a scientific truth but also a principle that has fascinated researchers and thinkers for centuries.

Digital Wearables, Movement, Psychology, and Exercise

As noted already, we are creatures who are designed to move, and we even experience joy and positive emotions when we exercise movement. Exercise is first on many health experts' list of priorities, yet it bears a certain stigma or a feeling of elitism. How did a shift to a more sedentary lifestyle become the norm, especially with the rise in digital technology that could help guide our movements?

I'm a technology designer who believes strongly in the benefits of exercise, and I see tech as supporting the drive. It's about the approach. It's not "all or nothing thinking." We can explore new ways of exercising that change the stories we might be carrying about ourselves. We can incorporate extended reality (XR) in the model because it increases flow, encouraging us to have less attachment to one fixed mindset, one limiting label, or one self-appraisal. We can all self-identify as athletes, and we can all find safer ways to train and support our bodies and minds for the long run.

Speaking of the long run, Sarah Meyer Tapia, director of Stanford Living Education (SLED), recently completed an in-depth study of the relationship between long distance running and digital device use. Tapia's research focuses on the field of mind-body wellness, holistic health, and the use of wearables in promoting health. Her doctoral thesis, "Running Naked," states the following:

> *Some use the term "running naked" with a tone of warmth and freedom, others with a lostness, a despondency. This study explored why. What is the lived experience of running with and without devices?*[11]

The study integrates exercise and mental health, wearable technology for sport, digital wellness, and mindfulness, flow, and transpersonal experiences in sport. Descriptions of device-free running were largely positive, referencing freedom, creativity, and connection to one's body, thoughts, and the running environment. Flow and transpersonal experiences were reported only while running without devices, with the notable exception of music. The intention of the study was to better understand how the presence or absence of devices impacts a runner's experience.

As I reflect on Tapia's findings, I think of how individuals feel over time about their habits and choices. Research findings offer a lot of insight into the potential for wearables and digital devices to interrupt the body's natural connected state. Among what stands out from the study takeaways are the often siloed nature of topics that need to be evaluated in context: "literature on mental health and exercise has not addressed the presence or absence of devices during exercise; wearable technology literature has rarely addressed the experience of the athlete wearing it; digital wellness literature has only just begun to include wearable technology for exercise; literature on flow states and transcendent experiences in sport has rarely mentioned whether devices were present on the athletes having those experiences." Incredible that the research has been so separated in the past.

I sense this is an important time to dig in and drill down on how we form meaningful connection with and without the augmented use of digital devices. Both sides have benefits and drawbacks on when it comes to choosing to combine digital wearables and running. Could Tapia's study indicate larger trends? It's my understanding that movement practices using XR technology can become a vehicle for transpersonal experiences that connect to flow. There is more to discover here.

Many people want to conflate topics of embodiment, XR technology, spatial computing, and AI, which can serve to support the immersive tech environments.

We exist in layers on top of our physical form. Our physical connection, and everything to do with those Seven ThEmes of spatial computing, comes first as a human-centered approach, with tech as a set of layers augmenting that human base.

We are biological first, and our wellbeing depends upon our understanding of that. This is why all of the beads on the necklace were designed with care and intention in this book, because they all connect and are joined by the thread of our humanity. Our understandings have formed layers of reflection and development through these chapters and discussions. Our topics are interrelated. We will be discussing AI in the next chapter because this very bead on the necklace is intrinsically connected to that one, with tremendous implications. I respond by encouraging people to disentangle topics. To look at overall life design holistically, yes, and to also understand the differences and distinct relationship between XR and physicality, between movement and embodiment, between AI and human. Think about objectives, coupled with what's emergent by staying present in the moment, tuning to your own extended awareness and state of curious, wonder-rich flow. Then, we can map intentions that will align with attention mindfully and lead to embodied, empowered action. Now is the time to be intentional, fluid, clear, and conscious about AI as we look to potentially integrate AI-driven processes into human-centered experiences.

Personal Reflections

1. What experiences do you have using wearables? How does using them feel? By contrast, how do your experiences without using a wearable feel?
2. How do you naturally respond to the phrase "body electric" and what ideas or wonderings do you have in response to the podcast and study?
3. What is your general level of movement during the day? Do you label this movement as exercise?
4. Do you have any experiences with sleep, meditation, or other restful activities while using wearables? If so, how does using a device feel?
5. Have you ever noticed your behavior change due to the Hawthorne effect? What have you noticed, and how does it affect what you prioritize in terms of maintaining authenticity?

Reflections for Leaders

1. How do wearables or any sort of biosensing play a part in your workday? Have you used any wearables for focus on energy monitoring?
2. Which of the five "body electric" truths and principles could you imagine implementing yourself or introducing to your team?
3. How much or how little do you think our corporate worlds honor the health of our physical bodies? How can digital wellbeing play a role in helping to support wholeness and thriving in ways that align with our business goals?
4. In what ways could XR and 3D interactive workspaces support the ideas of spatial and physical wellbeing?
5. Using a design thinking exercise, which is framed by the language, "I used to feel ______, and now I feel ______," what views that you might have had about the topics of wellbeing through wearables and movement have shifted or changed through the course of this chapter?

Extensions for Further Exploration

This chapter on wearables and movement is closely linked to music. Every topic is linked, in fact, as connected chapter beads on the necklace of wellbeing enlivened by wonder. I'm interested in how we increase freedom of movement and joy that we find in exercise at younger ages to translate to freedom of movement throughout the life-span, without pain, and yet with safe challenges and a feeling of accomplishment that embodies dignity for everyone. I meet many people who wish they could move "like they used to" at younger ages. I also know how much emphasis certain cultures place on organized sports, some of which seem less healthy for the body over time, due to impact and sustained injuries. That said, I'd like to incorporate more study of kinesiology into digital wellbeing and show ways that digital tools can encourage embodied experience—back to spatial XR themes!—as well as knowledge about ways that sports and ergonomic movements can enhance thriving. There's more to explore here.

I would love to curiously investigate more in the arena of music, movement, and meditation. Using wearables for biofeedback and music integration is of special interest to me; tracking in numbers and an interface that tells you how many steps to take a day are less intriguing, though

I am interested in how to motivate and positively change someone's story about themselves. I'm also interested in how wearable development can be more lifestyle friendly, nonaddictive, accessible, and suited to a wider diversity of population. We need to think in more ways that are JEDI: just, equitable, diverse, and inclusive in our digital wellbeing applications. I would like to explore that, with wonder and awe as conduits.

11 Wellbeing Through AI and How to Thrive with Hope

As we look back on the journey of this book, we recognize every bead on the necklace as part of a connected whole. A thread unites all of the topics we have explored. That thread is our humanity, our enduring sense of wonder, our awareness of the present moment, and the intention we place on directing our attention in ways that maximize our life thriving as an interconnected system. No one is isolated, and no one is alone. There is a sense of hope and tremendous possibility when we remind ourselves of that thread.

We are also mindful of the precious, vulnerable state of the world right now. From climate change to global political affairs, there is volatility and turmoil. There is a lot to attend to, and now is a time when diligence and conscientiousness can make all the difference in laying groundwork. It's a critical time right now to think about AI as its arrival in mainstream media comes on the heels of what Tristan Harris calls *The Social Dilemma*. AI might not be considered "media" in the same way "social media" is media, yet I call it a form of media because it is an intermediary, serving to bridge divides and ultimately mediate a human-to-human connection. This is not the only connection it can serve, yet it is a vital one. With AI especially, there will be serious consequences if we do not realize the power that's at our digital fingertips right now.

This chapter, this bead on the necklace, does not attempt to address all of my views about AI. There is a lot to share, and in the scope of our focus on wonder-rich digital wellbeing empowered by imagination, we will explore some of the ways AI is shifting our cultural narrative, talk about the connections between AI and our digital-physical lives, and address some ethical principles and takeaways that are core to our sense of resilience, dignity, agency, and humanity in an often AI-mediated landscape. There has never been a more wonderful time to be human, with so much possibility. And, as we have illuminated, so much is at stake.

We're entering an age of even more remote work, even greater distribution, and rapid rises in complexity. Which brings me to Iwo Szapar, whom I first met in person at the LinkedIn annual leadership summit in Los Angeles, where I was giving a workshop. Szapar and I were in a session about the future of remote work. We started talking about opportunities in AI and XR to ignite imagination and freedom. Right away, I was struck by his genuine enthusiasm. He already seemed to be living in a future scenario, untethered by some of the vestigial outdated architectures of society.

As cofounder of Remote-how, a company focused on enabling remote workforces and travel, Szapar has co-created initiatives including Virtual Coworking and Remote-how Academy, and his community comprises more than 25,000 members from 128 countries. His ideas on AI and the future of work revolve around the impact of AI on distributed and remote workforces. He believes AI is a catalyst for the distributed work movement, making work not just easier but better for distributed teams.

Many people feel threatened and unsettled by this rapid onset of AI. According to Szapar, AI might be aptly perceived as disruptive, but it is a benefit rather than a threat to workers. It needs to be viewed and adopted as a set of tools that can empower the workforce by automating best practices and optimizing workflows. This involves a tremendous mindset shift for much of the global population. The projected growth of the AI market to $4 trillion by 2025 underscores the increasing importance of these technologies in the ever-evolving workplace. Furthermore, AI has the potential to enhance the knowledge worker experience, with the possibility of companies developing their own large language models (LLMs) that can learn and improve the operations of individuals and teams. Additionally, the integration of AI into the workplace could lead to a significant shift in work-life balance, potentially reducing the work week to just three days according to some predictions.

These harbingers and ideas signaling major changes on the horizon are backed by reports and market projections, such as the 2024 McKinsey report on the AI market,[1] the 2024 World Economic Forum publications on AI and the future of jobs,[2] and the AI Index Report published by Stanford's Institute for Human-Centered Artificial Intelligence.[3]

Many call 2023 and 2024 "breakout years" for AI, due to its rise of prevalence in the working world, also spilling into nearly every area of life. Szapar and I discussed the application of AI in enhancing remote work readiness and performance, providing personalized recommendations in myriad languages, and offering cost-effective insights about improving workflow and operations. Much of the larger discourse about AI focuses on the evolving nature of work in the context of technological advancements. I asked Szapar to comment on his daily life and outlook, with AI impact in focus. Here, he shares views about AI and the future of work.

AI's Impact on Daily Routines

AI is enabling us to look at our quotidian rituals differently. Szapar says, "AI has fundamentally transformed my daily routine, making it significantly more efficient. By automating 80% of my tasks, AI has become an indispensable brainstorming partner. This technology has not only streamlined repetitive tasks, but has also enabled me to adopt best practices almost effortlessly. The result? My days are now structured around leveraging AI's capabilities to enhance productivity rather than being bogged down by monotonous tasks. This shift has allowed me to focus more on strategic thinking and creative problem-solving, fundamentally changing the nature of my work for the better."

Advice on Using AI for Remote Work

AI can help humans stay more connected, more adaptive to changes, and even more responsive to each other. It can also increase our playfulness, as we use it in specific ways that enhance our own creative freedom. "In the realm of work, particularly remote work, integrating AI into daily operations is not just a recommendation—it's a game-changer," Szapar says.

"My advice is straightforward: use AI daily and embrace experimentation. The power of AI goes beyond merely speeding up tasks; it enables the implementation of effective work rituals that forward-thinking professionals have advocated for years. By harnessing AI's capabilities, you can enhance not only the efficiency of remote work but also its quality, paving the way for a more dynamic and responsive work environment."

These points go hand-in-hand with our earlier chapters about imagination-rich learning systems and immersive spatial environments and experiences, powered by AI so that we can focus on empowering and supporting our ideas, our connection, and each other.

Benchmarking Success in an AI-Enhanced Future

When Szapar and I talk about the future, we have bright outlooks. They are also aligned with new assessment models, and new key performance indicators (KPIs) that are clearly needed in order to measure this next wave of flourishing in business and—most importantly—overall life. "As AI takes on the heavy lifting in terms of productivity, our benchmarks for success need to evolve. While 'wellbeing' might seem like a broad term, it captures the essence of what truly matters—personal satisfaction that extends beyond professional achievements. To measure the impact of AI on our lives, regularly conducting employee Net Promoter Scores (eNPS) can provide insight into how technology influences our wellbeing. The ultimate goal is to create a future where AI's support in productivity leads us to work less yet achieve a higher state of happiness and fulfillment. This approach ensures that as we advance technologically, we don't lose sight of what makes us human—our quest for satisfaction and wellbeing."

Creativity, wonder, and imagination are clearly part of the framework, as we evolve to think bigger about how AI can help us all achieve fuller lives of thriving, at global scale. A commitment to wellbeing through improving quality of work life is evident in Szapar's outlook, and he's an example of someone who has maintained a disciplined lifestyle in order to meet professional and personal goals and dreams, which include wellbeing priorities like surfing and travel, being able to attend global conferences, meet new friends, and still carve out time at home with loved ones. Our conversations are about business, workflow, and our mutual projections for the future, and we are conscious of how AI offers opportunities to enhance quality of life.

This AI moment, and decisions we make about how to thrive, make such a difference.

AI, Wellbeing, Social Media, and What's at Stake

AI development is urging us as individuals and as a global society to make a stand in support of our broad wellbeing, which is why there's a lot at stake. I do not want to seem dire, but the time to act and establish guidelines and policies is right now.

Learning from the past, three main subjects should be considered as we make choices about our next actions regarding AI:

What are the situational commonalities between social media and AI?
What mistakes were made with social media?
As a corollary, what could—and should—we do differently?

Social media and AI have many parallels, including their popularity. When social media was first on the rise a couple of decades ago, it was seen as a means for connection, information sharing, and empowerment. AI bears the same promise. Both have the potential to make a positive difference in relationships and to increase wellbeing through making life more efficient, vibrant, and connected.

What does this connection supported by AI truly mean, though? Is it noise or meaningful content? Are people connected and whole internally as they seek external connection? Are they connected together or unwittingly split into silos of polarization and distracted by disinformation? And is the information—media—that they consume trustworthy? We cannot leave it up to individuals to judge the veracity of what they consume—imagine if we approached the food industry that way. All of this depends on standards and regulations to ensure quality. Our solutions lie in responsible limits, education, and relational trust.

In addressing the mistakes made with social media, we also address the possibility to attend to those mistakes and become more aware of them. Our greatest asset, after all, is our attention, which is how we spend our time. Where we place our attention matters, showing our intention. Social media found ways to hijack our attention, keeping us hooked and wanting more in a game that Harris has called, "the race to the bottom of the brain stem." We were treated like beasts in the equation rather than empowered to make our own choices. Various tricks to trigger dopamine and serotonin have been used, and the experiences interacting with social media have generally been unfulfilling rather than uplifting for most people. So why do we continue to use it? Because of our need for connection and the promise that it bears.

In an article in the *MIT Technology Review* about social media and AI, Nathan E. Sanders and Bruce Schneier identify five fundamental attributes of social media that have harmed society: advertising, surveillance, virality, lock-in, and monopolization.[4] These same five categories have the potential to be massively mishandled when it comes to AI integrations. If we approach AI in the same way as social media, targeting user engagement, mishandling data, and favoring big tech profits, the impact at global scale—societally, environmentally, economically, across all sectors—will be exponentially worse.

AI, Digital Ethics, and Data Dignity

"Data dignity" is a term being brought up all the time these days, and it was coined by Jaron Lanier, a well-known tech commentator and virtual reality creator, before the surge of AI hit our cultural mainstream. Data dignity is a concept that advocates for a shift in the economic model to ensure that individuals are compensated when their personal data are used or when data they have created are remixed. This movement emerged before the advent of generative AI and is rooted in the belief that the current "free" online economy has failed to adequately recognize or remunerate individuals for their contributions. The introduction of generative AI is expected to exacerbate these issues, making the need for data dignity even more pressing.

The concept of data dignity also addresses concerns about the metadata associated with information—the "information about that information." Lanier and others advocate for a system akin to a "baggage tag" for information to ensure that metadata are not lost as data travel across the web. This is crucial because, without such measures, there is no straightforward automated method to verify the authorship of a document or ensure that quotes are correctly attributed if they appear elsewhere. This lack of attribution is particularly problematic as generative AI technologies, like GPT-style AIs, have the capability to aggregate vast amounts of user data from the web, often without preserving the origins of that data. The convergence of these issues—misuse of personal data by large corporations, the capabilities of new AI technologies, and the absence of a system to track authorship digitally—highlights the urgent need to rethink how we interact with and manage data on the web.

Now, imagine that data are not only on 2D interfaces, but also 3D spatial computing interfaces, and imagine it related to our biometrics. Where is our data sovereignty? There are issues with our rights to protect our data across these interfaces, our need to have awareness about the way the data are shared and where they are stored, and our rights to compensation for our attention.

There are a few possible ways that this situation could be helped. The Semantic Web concept, introduced in the 2000s, aimed to enhance the World Wide Web by enabling intelligent agents to understand webpage content through injected metadata for improved human interaction and service provision. However, the Semantic Web project faced challenges due to complex metadata languages, slow inference engines, and the evolving nature of metadata. In the context of search engine optimization (SEO), metadata often served commercial interests rather than enhancing truth and accuracy. We need to have better plans in place for AI.

There are several possible solutions, including using AI to generate and maintain metadata, ensuring accuracy and relevance in a dynamic digital landscape. AI could automate the creation and upkeep of metadata, enhancing data integrity and attribution. Also, we could use improved tools and standardized agreements to reintroduce consistent and valuable metadata into the web ecosystem. By establishing clear guidelines and frameworks, we can ensure the quality and relevance of metadata across digital platforms.

Storing metadata in the cloud is also a way to facilitate accessibility, scalability, and maintenance. By centralizing metadata in cloud environments, we can streamline data management processes and ensure its longevity and accuracy. These ideas are not perfect solutions, yet they would let us start somewhere. The key will also be in the interoperability of AI systems and ethical standards that keep everyone involved prioritizing taking action. These actions include diversifying data sets, checking source reliability, and taking into account broad perspectives so that the AI models are trained on a quality range of inputs. This is part of how we can approach digital wellbeing and AI to ensure better thriving.

Risks and Rewards of AI

While we acknowledge opportunities and rewards, complex challenges and potential risks are associated with the development and deployment of AI. One of the major concerns is the possibility that humans may lose control over AI, which could lead to unintended and potentially catastrophic consequences. Harris has highlighted that half of AI researchers believe there's a significant chance that AI could lead to human extinction if not properly controlled.[5]

The Three Rules of Humane Tech

Tristan Harris and Aza Raskin, co-founders of the Center for Humane Technology (CHT), have proposed three rules of humane technology to guide responsible technology development in the age of AI.[6]

- **RULE 1: When we invent a new technology, we uncover a new class of responsibility.** We did not need the right to be forgotten until computers could remember us forever, and we did not need the right to privacy in our laws until cameras were mass-produced. As we move into an age where technology could destroy the world so much faster than our responsibilities could catch up, it's no longer okay to say it's someone else's job to define what responsibility means.
- **RULE 2: If that new technology confers power, it will start a race.** Humane technologists are aware of the arms races their creations could set off before those creations run away from them—and they notice and think about the ways their new work could confer power.
- **RULE 3: If we do not coordinate, the race will end in tragedy.** No one company or actor can solve these systemic problems alone. When it comes to AI, developers wrongly believe it would be impossible to sit down with cohorts at different companies to work on hammering out how to move at the pace of getting this right—for all our sakes.

As a whole, the three rules of humane technology are all interconnected, highlighting the ethical challenges posed by modern technology and the importance of designing and using technology in ways that prioritize human wellbeing and societal good. They emphasize the need for coordination and cooperation as humans, which are skills associated with wellbeing, wonder, and imagination at core, to see from multiple angles and embrace possibility. This advocacy is part of a much-needed conversation about the ethical use of technology in our society.

What I appreciate about these dialogues and proposed solutions is their consideration for how we think about our own thinking and decision-making. The best way to approach change is through curiosity, and Harris does this, speaking about very serious, dire issues with candor and openness. One post about 3D thinking also caught my eye, especially given the focus of this book and our constant reminder to ourselves that we inhabit 3D bodies and live in a spatialized world.

Technology reflects intentional design. In a 2023 *TIME* article addressing the flawed 2D design choices of the past that prioritized engagement as a business model, Harris says we can make new design choices: "We can rebuild existing technology with a more humane philosophy that strengthens our capacities for synthesis and cooperation. Change won't be easy, but the cooperation we need to tackle all of our other existential challenges is at stake."[7]

Three-dimensional technology in the world of spatial computing and this book's multifaceted exploration have shown many pathways to digital wellbeing that have new priorities in place, new KPIs, and new metrics for thriving. Through this book, we have guidelines, mindsets, and frameworks for how to approach an AI-powered future.

> *Three-dimensional technology in the world of spatial computing exploration has shown many pathways to digital wellbeing that have new priorities in place, and will elevate new KPIs, and new metrics for thriving.*

These new ways also require a process of letting go. Embracing change means looking at our own shadow sides and being willing to see the imperfections. It also requires asking ourselves why we are engaging with technology or employing the use of AI in the first place. The word "manipulate" comes from the same root word as "hands." Are our hands free-willed and able to manipulate our own experience, or are we being manipulated? Could it be both? How can we become more 3D-minded about these issues? How do we embody our values and use somatic practices to tune into the physical, as we layer the digital? These are among our considerations, and there are many more questions and still a great chance for a positive outcome filled with agency and wellbeing.

AI-driven Solutions for Digital Wellbeing

In sum, there are specific ways that AI could empower better, wonderful, awe-inspired digital wellbeing for all of us, while we maintain consciousness about our choices and overall intention. This is a possibility, not a mandate. Feel how these possibilities land for you, and imagine if and how they would resonate with your life. We have seen many of these related topical beads on the necklace in past chapters. For every point here where AI is possibly integrated as a solution, imagine if and how you would want AI to affect your life in these ways.

In areas of personal recommendations, AI can analyze data from wearable devices to provide us with personalized insights and recommendations for improving our physical health, mental health, and sleep. While this can empower individuals to make more informed decisions about their digital wellbeing, I've found that I often prefer to make my own choices. Still, the feedback from these devices sometimes uncovers what I could not have predicted. And I think, surprised, "that's interesting." My awareness rises. I make more informed decisions, if there is trust in the inputs and dignity defended.

Imagine, AI-powered wearables can track physiological parameters we mentioned earlier, including heart rate variability, skin conductance, and breathing patterns to detect and monitor users' mood states, stress levels, and emotional wellbeing. AI-based digital assistants can be designed to monitor user behavior, detect unhealthy patterns, and provide tailored recommendations to promote healthier digital habits, such as screen time management, social media usage, and work-life balance.

As we mentioned earlier, AI can be leveraged to enforce ethical data practices, ensuring secure data management, transparent data usage policies, and user-centric control over personal information, thereby enhancing digital wellbeing and data dignity.

In addition, we could use AI to generate an overall "map of our wellbeing," enabling it to integrate data from various sources, including wearables, smartphones, and environmental sensors, to provide a more comprehensive understanding of our overall wellbeing, enabling more holistic interventions and support that align with our physical, mental, and emotional wellbeing.

On top of all of this, we can use AI to support our dreams for lives that are imaginative and filled with freedom, wonder, and awe. We can exist in healthy, connected relationships with ourselves and others, focusing on what lifts us up.

By incorporating these AI-driven approaches, the digital ecosystem can be transformed to better support individual and community digital wellbeing, foster transparency and accountability, and empower users to make more informed decisions about their online activities and data. This, of course, all depends on our guiding principles and ensuring that diversity and accessibility is honored. Back to our Four Culture Cornerstones, now is the time to remind ourselves of the dignity, freedom, invention, and agency we must uphold, using AI to support us and not to overwhelm us or strip us of our sovereignty. How we thrive with hope involves staying conscious of our values, asking informed questions, and being deliberate about the ways we develop and use AI, as individuals and as a global community.

Personal Reflections

1. What brings you back to yourself when you feel depleted? Do you think AI can help with this, giving you more insights?
2. What social media or AI tools give you a lift? In what way? Are you more or less energetic afterward?
3. How do you see your relationship to AI? Do you envision it as a smart colleague? A talking book? A navigator or copilot? Do you ever find yourself anthropomorphizing it, imagining it has human traits?
4. How has your life been busier or less busy in the last two years, since the rise of generative AI? How can we meet this moment from a digital wellbeing standpoint?
5. What do you feel your personal world is asking for, and what is the larger world asking for right now? How do your choices about how to incorporate AI into your life reflect what these worlds are asking for, and could AI help you respond with even more resilience?

Reflections for Leaders

1. When it comes to AI, how do you feel about the adage "Automate the routine so you can humanize the exceptional"? Along those lines, if you could delegate the most mundane and repetitive tasks in your day, what would they be? What would that free you up to engage in? How would that benefit your life and business by expanding what you do?
2. Imagine that an AI twin was created related to you and your organization. What values would you like it to reflect? What details and materials would you feel comfortable giving to train it?
3. When you use new AI tools, do you feel lighter for the workload lifted? Is there also a feeling of heaviness for having to learn new software and methods? How do you respond to these feelings?
4. As we are in the midst of a tremendous job transformation, how are you experimenting and getting playful in your approaches to AI, looking to extend your capabilities to compete in these new job markets, and to support your colleagues and teams? How does a wonder-rich digital wellbeing mindset, which we have built throughout this book, help you in your adaptation and curiosity?
5. How do clarity and experimentation meld and merge with safety and regulation when it comes to AI in our organizations? How can we as leaders encourage sandboxing and innovation, and learn more about this emergent technology in order to have benefits and not harms?

Extensions for Future Exploration

There are many expansions and explorations, too many to list here. I will share even more research and views online, where you can follow updates. Among many curiosities and passions I have is using AI in applications involving empathic computing and imagination. One of the top considerations, especially with the global climate crisis, political instability, and growing rates of depression and loneliness, is that it's important to increase public debate about how and why to use the vast amounts of AI tools and applications available to us. Now is the best time to establish standards, infuse awe, and redefine relationships. Digital wellbeing through a lens of wonder is paramount, and will transform how we approach AI with goals of full-life thriving.

Future Sight: Envisioning a Future of Digital Thriving

"When you don't name things anymore, you start seeing them."

—Alan Watts

"Words are our weakest hold on the world."

—Alberto Ríos

"To name something truly is to lay bare what may be brutal or corrupt—or important or possible—and key to the work of changing the world is changing the way we can imagine it."

—Ursula K. LeGuin

While this book has done its fair share of naming and demystifying digital terminology, its purposes are beyond words themselves, which merely signal what invites us into experience.

This book has been a place to practice the art of naming and seeking to understand what is complex and emergent, all in the name of context-driven digital wellbeing through a lens of wonder. The concepts in each chapter have been related to each other and also independent, which is why we've used the metaphor of beads on a necklace to remind us of interrelatedness, connection, and individuality. Whether we are talking about social media or virtual reality, exercise and movement, or mindfulness and wonder, we are talking essentially about relationships.

It is our relationships—with each other, with nature, and with ourselves—that are mediated by the technology we choose to adopt, for personal use or within our businesses and working worlds. The tech itself is layered and nuanced, and it also can have emotional resonance. At each stage and step, with each bead, we also acknowledge that there is a relationship with the tech itself to consider. This book favors consideration over easy, prescriptive answers that are tempting because of their simplicity, yet ultimately dissatisfying because of their false promise. The reality is situational, and I have trusted, throughout this book, in my own intention to place my attention on *you*, first and foremost.

What could I possibly mean by that? The reality of this book is that it is a transmission from author to reader, and everything is media. Everything is a transfer. I wanted to write this book in a different way, eschewing conventions and focusing on presence, making my presence palpable to you and doing my best to listen to the world and to focus those curiosity-driven findings into each bead, each topic, which then has its own integrity and will become a space from which to grow. Through this growth, we can elevate our ways of being and all reach states of digital wellbeing that work best for us.

This has been a journey that is meant for your thriving with technology. Each chapter has reflections, stories, discoveries, and examples we can each use in daily life. We started with a base of mindfulness and presence, and we have used imaginative practices in wonder throughout as our guideposts. While I've given the frameworks that serve as guides and supports, this is an invitation for you to apply those methodologies. Throughout, you are encouraged to feel grounded and supported by your own experiences and your own investigations. You can apply this journey to your work life and your personal life, to your teams and to yourself as an individual. It is accessible and filled with ideas and learnings that are intended to spark your imagination and sense of wonder.

As we close this book's exploration and look forward to the ever-expansive future of digital wellbeing, here are five lenses for our Future Sight, to add to our primary lens of wonder, which we now have as part of our vision. Lenses can be layered.

Rather than predictions, because who knows what's coming next and I would rather not prognosticate, here are five views that I believe will help guide this next phase:

Lens One: Praise and Practice Your Plasticity

Our brains are adaptive and curious, ready to predict the future by comparing it to the past, encoding situations with emotion and meaning. One of the reasons this is important to our future digital wellbeing is that we can use this knowledge to try to "think outside the box" as much as possible. By taking advantage of our brain's plasticity through practices in daily mindfulness and in acts of noticing, we flex our "wonder and awe muscles," which has a direct effect on our quality of experiences and way of showing up in the world.

We can practice more intentional acts that strengthen our resilience, as we talked about earlier in this book, and by focusing on adaptation, we give ourselves more opportunities to stretch and grow as individual humans, and we also raise the bar for our whole species, thinking more creatively and open-mindedly. This translates to better whole lives: personal, educational, business, societal, planetary. It's better for everything and everyone. With the intention to widen the lens, we stop thinking of ourselves as individual, and we become more collective minded.

Because it was built to enhance states of open-mindedness, creativity, and flow, extended reality applications are ideal spaces to practice this plasticity, as we have explored earlier. The quality of platforms, hardware, and software will keep improving, and now is the ideal time to set sights on this future-forward form of connecting, learning, and exploring.

Lens Two: Digital Wellbeing as a Way of Being in Right Relationship

The future of digital wellbeing is asking us not to ignore our physical selves or leave our biological forms behind in some sort of transcendence, but to layer digital wellbeing in a way that is human as we approach technology.

Just as mindfulness offers a way to be in the world—there are new ways to approach collective digital wellbeing. It's the "being" that counts. We are relational creatures, and our attention, social bonds, and formation of relational trust is important. What is a "right relationship" with ourselves and with each other?

Many of us notice how attention on smartphones and other digital devices have served to divide relationships. It doesn't have to be this way, and viewing digital as the pure enemy is in ways too simplified. There are ways that technology can bring people together, in better understanding, with more curiosity and wonder. I see this as a future lens: How can we increasingly use digital media as a bridge rather than a divide? The connection needs to be internal as well as external. Connect inside first, then connect outward. Don't replace physical. Augment it. Integrate it.

We're never going "back" to the way things were . . . that's an impossibility anyway. Change is constant, and the recent digital past has been a form of a revolution in terms of expectations and human connection.

We're at a pivotal moment here, right now, one that prompts inner and outer reflection. *Take your time.* Honor the questions and the process of listening to each other. Put your phones and devices away in the process of listening. I predict that in the near future it will be seen as rude and insensitive to have a phone out at the same time that we are having a physical intimate conversation. Am I ambitious in saying that? Could we make it a societal movement? After all, we are the ones who can harness the technology to serve us well. Let's ask it to serve us.

This is our right relationship, intention-meets-attention digital wellbeing moment, right now. This is where it all starts. Celebrate the art of getting to know the person next to you. Dig in. Most of all, dig into that relationship you can have with yourself. It endures.

Lens Three: AI Is Even More Momentous than the Hype

Machine learning is not new; it's just the mainstream adoption that has changed the scope of our attention on AI. Most of AI use at this stage is most beneficial when incorporating it as an assistant, a helper tool, or a copilot. It can organize and automate.

Could it become a more joyful experience to use AI? Certainly it's efficient, as we have seen in this book. Many people—not just the creators or the technologists—need to feel invited to learn and understand more about the power of AI. As Sal Khan has said about artificial intelligence, "We all have to become literate. This is different than anything that's come before." There's wisdom at stake here, as becoming literate takes time, and the data sets need diversity and discipline. That and data diligence are top of mind.

Hearing founders and experts speak openly about all the questions that are still unanswered about this new technology, inviting our help, is promising. We live in fascinating times. Extended learning communities engaging about these topics and shedding light on the critical nature of this epoch inspires me to jump in and participate in dialogue, even and especially when each of us alone doesn't have all of the answers. Because we are each helping to shape the conversation and our future, it feels critical. I sense that our future of wellbeing and thriving depends upon this cooperative, collaborative, diversified dialogue. All of our voices are needed, and deep listening is also necessary.

Lens Four: New Key Performance Indicators Are Social and Emotional Wellbeing

Empathy is going to reframe the future, as social-emotional connection with other humans will be increasingly important. More learning can happen autodidactically in the future, with help of AI tools and multigenerational teachers and mentors. In my vision of an elevated future, everyone in learning systems will be compensated fairly for contributions so that these systems of learning are ones of collective abundance instead of scarcity. Since all institutions depend on quality education and opportunity, this will improve access to quality health care and a reframing of societal infrastructure. If AI helps us to improve efficiency and automation, we will have more time to focus on quality and more abundance of what matters. Financial wealth is currency—again, a form of media. What will it serve to connect? The new key performance indicators (KPIs) are going to be centered on quality elements of thriving, which all center on wellbeing. Healthy living, healthy dying, and social-emotional impact will be important topics all influencing wellbeing. The digital aspect of wellbeing is simply a set of media.

AI tools empowering social emotional interactions, proposing better wellbeing incorporated into learning design for all ages, will help shape a

We will better learn how to learn better and how to better understand ourselves and our minds in the process.

bright future. Learning will evolve to encompass the entire life-span. We will cultivate capacities of our mindsets, and from there we layer on skillsets and the toolsets with which to accomplish our goals. We will learn how to integrate with machine learning and AI assistants. We will better learn how to learn better and how to better understand ourselves and our minds in the process. Humans are social animals, so this does not happen in a vacuum. This happens in a social context, as we grow and evolve over our human life-span, which can now be approached as a "health-span."

Digital wellbeing in this reframing opens us up to the preciousness of life and the possibility to reappraise approaches to longevity. How do we as a culture view death and dying? Are we intergenerational in our friendships, and how do we view the world? How do we view our own aging process? As we focus on living well, we could also look at what it means to be dying well. Our lens on social and emotional quality KPIs of wellbeing has a wide enough scope to include aging well and dying well. Many cultures worldwide view the process of aging and death as sacred, and the future will involve animating and connecting these traditions as we access greater states of wonder and imagination about what a life of thriving in connection can be.

As we reshape approaches to our communities and our health-span with more intentionality, one consideration will be how we can prime for conditions that will make our freedom of choice surrounding life, love, longevity, and death easier to honor and access across all socioeconomic levels, in all areas of society. How will we form new literacies for long-life learning and celebrating a full life-span and health-span without denying the sanctity and joy that end-of-life rituals and practices can encompass? It starts by dialoguing about it and imagining what we want for ourselves and those we love.

Lens Five: The Invisible Will Become Visible and Beautiful to Behold

Visionary ideas have the ability to "flock," as birds do, to cosense in live time, and to take flight. Patterns can become metaphors. Networks and arrays in many forms and across disciplines are emergent themes. Nature and its powerful way of communicating and cosensing can increasingly become a teacher and metaphor for us, as we look to design well in the future, with AI to help us "decode" and understand even more about the beauty, wonder, and wisdom that surround us and are within us. We are a part of everything, with wonder and awe as guides.

The coming epoch in human life could be one that focuses on beauty. Not superficial beauty—deeper beauty. Our human appreciation for what moves us, for what sparks wonder and inspires awe, will be what we value. Our priorities will encompass moral beauty, and entertainment will expand beyond superficial engagement and dopamine spikes as we incentivize for wellbeing and authentic connection in a collaborative network.

One prime example is fungi, with its powerful network structure and interconnected distribution of nutrients. Scientists estimate that, while 100,000 species of fungi have been identified, there are 0.8–3.8 million species on (and in) the planet. Fungi are one of earth's biggest recyclers, breaking down compounds and turning them into nutrients available to enable the regeneration of life. As a result, they are major drivers of soil health and carbon sequestration. They are models of resilience and cooperation, evident in their mycelium networks, adaptability, and symbiotic relationships. We can learn a lot from them and apply these principles to our behavior, society, and culture in ways that enhance physical-digital wellbeing.

Another example are Aspen trees. An Aspen stand of trees is in fact a giant clone network.[1] They are a singular organism with an extensive root system. They have the ability to share resources, cosense need, and regenerate when parts are dying off.

A Stanford study found that "working collectively on a task can supercharge performance."[2] In the future, it will be imperative to encourage that collective mindset and allow people more chances to see themselves as part of a greater whole. In a time of loneliness and isolation, this message is empowering and needed for sure! It's a Future Sight outlook that we will look to the beauty and resilience of nature and use these models for our own thriving in the future.

Many forms of art can inspire us and be a gateway into this collective mindset. Doris Mistch's animated photographic peregrinations show us the flight patterns of birds in ways that are emotionally moving. Across space and time, flight patterns became fresh translations of motion. Movement became meditation, as we witnessed and experienced a new way of seeing wings, and consequently we can allow our own ideas to become expansive.

Researcher Karen Bakker has talked about the hidden sounds of nature, showcasing high ultrasound and deep infrasound. Just because we don't hear it, doesn't mean it doesn't exist. We look for proof through our sensory experience, and yet there's an entirely separate world hidden in plain sight, and technology can make it apparent to us. The singing of whales can leave us rapt.

AI is now helping us with pattern recognition and translations, allowing us to better understand the world around us. That scientific understanding is also an emotional and moral understanding, encompassing an inherent beauty. As Dacher Keltner has pointed out, recognizing moral beauty as well as the awe of nature are the second and first most prevalent forms of awe for humans. These forms of awe elevate belonging. They do not have to involve another human (though they could); these awe states are quite moving for us to experience, and they place the surrounding world in a certain context, a critical significance that informs how we live and what moments we prioritize.

The future will involve, once again, prioritizing awe like never before, in part because that is how we create meaning for ourselves. AI can be delightful, helping us to see what was unseen, find delight in new ways of approaching data, and form deeper appreciation for the world.

For example, when Refik Anadol creates an AI art installation, he uses large datasets as the raw material for his creations, employing machine learning algorithms, particularly Generative Adversarial Networks (GANs), to interpret and visualize data in innovative ways, creating what he refers to as "data paintings" and "data sculptures."

Standing in front of Anadol's art, I'm witness to the other humans reacting to the space. His AI-generated art moves and evolves in dynamism with the physical world. Each time, it's the crowd's collective effervescence (a powerful form of awe) in cosensing and sharing the experience that makes it unique and revelatory as we interact with it.

Many people will say that the greatest commodity we have in life is time. How we choose to spend it is our greatest gift. I would add that ***time is* nothing *without our attention and intention placed on it.*** That is mindfulness, as we identified earlier, and we can apply it to digital wellbeing.

The time itself can drift by if we are in a mindless stupor. That said, the ability to hold attention loosely, to bask in awe and imaginative reverie, and to let go of a fixation on time itself—this is all part of the skill, nuance, and balance that wonder + mindfulness can give us. It's a way of reaching what some, including the researcher Mihaly Csikszentmihalyi, call flow state. It's a practice, even, to name emotions and be with ourselves in these moments, attending to the inner landscape, while appreciating that outer world and our belonging in it. Good friends make such a difference on this journey. The greatest lesson is self-love, expansively extending from that core.

Conclusion

As this exploration of digital wellbeing draws to a close, it's not without a sense of paradox. The act of writing has been a profound journey, a meditation on wonder through the digital lens that has brought forth a glow, an uplifting energy sparked by the many visionaries and innovators whose stories and insights fill these pages. My gratitude lies in the insight that each chapter here is the beginning of deeper dialogues to come. It's an expansive wisdom source, rippling out.

Chapter topics have all been beads on a metaphorical necklace, strung together with the thread of reflection, intuition, and humanity. We delved into the well of being well, contemplated the balance of attention, and sought understanding in the immersive interactions of spatial computing. Music's rhythm became a metaphor for the heart's desires, wearables a symbol of the body's dialogue with the digital world, and AI a reflection of our hopes and fears as we stand at the precipice of an evolving epoch.

This book is an invitation—a call to recognize that digital wellbeing is not a static ideal but a living, breathing relationship. It's a guide through the moments of impasse, offering strategies to navigate the ever-changing digital landscape. That is what I wish for you, too, in reading this book and putting the messages into action in the ways that speak to you and work well for you. Wellbeing, including digital wellbeing, is a dynamic construct. Importantly, we know that shifts are natural, that life is not a straight path, and that every ending can be the seed of a beginning, and from that an elevation in growth.

As you, the reader, reach the end of these pages, may you find not just the end of a discourse but the start of a personal revolution. May this book serve not only as a compendium of ideas but as a catalyst for exploration, drawing you into experiences that deepen your sense of belonging to this wondrous, interconnected world, and heighten your sense of freedom through imagination. Here's to the paths you will forge, the relationships you will deepen, and the strides you will make in the dynamism of wonder-rich wellbeing in the digital age.

Notes

Preface

1. LitCharts, "The Vinegar Tasters Symbol in the Tao of Pooh," n.d., https://www.litcharts.com/lit/the-tao-of-pooh/symbols/the-vinegar-tasters.

Introduction

1. Brené Brown, "Dr. Susan David: The Dangers of Toxic Positivity, Part 1 of 2," Brené Brown, March 1, 2021, https://brenebrown.com/podcast/brene-with-dr-susan-david-on-the-dangers-of-toxic-positivity-part-1-of-2.
2. NLI Staff, "Latest from the Lab: Regulate Your Emotions, with Help from a Friend," NeuroLeadership Institute, September 21, 2023, https://neuroleadership.com/your-brain-at-work/latest-from-the-lab-regulate-your-emotions-with-friends; Arasteh Gatchpazian, "Vulnerability: Definition & Tips," The Berkeley Well-Being Institute, n.d., https://www.berkeleywellbeing.com/vulnerability.html.
3. Gang Wu et al., "Understanding Resilience," *Frontiers in Behavioral Neuroscience* 7 (2013)., https://doi.org/10.3389/fnbeh.2013.00010.

Chapter 1

1. Kai Ruggeri, et al., "Well-being Is More Than Happiness and Life Satisfaction: A Multidimensional Analysis of 21 Countries, *Health and*

Quality of Life Outcomes 18, no. 1, https://doi.org/10.1186/s12955-020-01423-y.

2. Gemma Simons and David S. Baldwin, "A Critical Review of the Definition of 'Wellbeing' for Doctors and Their Patients in a Post Covid-19 Ira," *International Journal of Social Psychiatry* 67, no. 8: 984–991, https://doi.org/10.1177/00207640211032259.
3. Better Health Channel, "Wellbeing," Victoria Department of Health, n.d., https://www.betterhealth.vic.gov.au/health/healthyliving/wellbeing.
4. Jonathan H. Ohrt, Philip B. Clarke, and Abigail H. Conley, *Wellness Counseling: A Holistic Approach to Prevention and Intervention* (Alexandria, VA: American Counseling Association, 2019), https://psycnet.apa.org/record/2018-65274-000.
5. Student Wellness, "Dimensions of Wellness," Northern Kentucky University, n.d., https://inside.nku.edu/studentaffairs/departments/student-wellness/dimensions.html.
6. Tchiki David, "What Is Digital Well-Being?," *Psychology Today*, February 22, 2021, https://www.psychologytoday.com/us/blog/click-here-happiness/202102/what-is-digital-well-being.
7. David, "What Is Digital Well-Being?"
8. Vanden Abeele, Mariek M. P. (2020). Digital Wellbeing as a Dynamic Construct, *Communication Theory*. qtaa024, https://doi.org/10.1093/ct/qtaa024
9. Jacqueline Beauchere, "Introducing the Digital Well-Being Index," February 6, 2023, https://values.snap.com/news/safer-internet-day-2023.
10. Louise Hatem and Daniel Ker, "Going Digital Toolkit Note, No. 6, 'Measuring Well-being in the Digital Age'," ed. OECD and Going Digital, 2021, https://goingdigital.oecd.org/data/notes/No6_ToolkitNote_MeasuringWellbeing.pdf; OECD, "Comparing well-being in the digital age across OECD countries," in *How's Life in the Digital Age* (Paris: OECD, 2019), https://www.oecd-ilibrary.org/sites/9789264311800-5-en/index.html?itemId=%2Fcontent%2Fcomponent%2F9789264311800-5-en. https://www.youtube.com/wach?v=9-k6sIN-2K4.
11. Mayur, "Krzysztof Kieslowski's Cinema Lesson (Master Class in Film Direction)," YouTube, August 2, 2013, https://www.youtube.com/watch?v=9-k6sIN-2K4.

12. Carolyn Gregoire, "Harvard Stress Forum: Can We Use Technology to Become More Mindful?," *HuffPost* (blog), December 6, 2017, https://www.huffpost.com/entry/technology-mindfulness_b_2828562.
13. Gregoire, "Harvard Stress Forum."
14. Ellen J. Langer, *Counterclockwise: Mindful Health and the Power of Possibility* (New York: Ballantine Books, 2009).
15. "Mindfulness in the Age of Complexity," *Harvard Business Review*, March 2014, https://hbr.org/2014/03/mindfulness-in-the-age-of-complexity; The On Being Project, "Ellen Langer—Science of Mindlessness and Mindfulness," November 2, 2017, https://onbeing.org/programs/ellen-langer-science-of-mindlessness-and-mindfulness-nov2017.
16. Pursuit of Happiness, "Ellen Langer," n.d., https://www.pursuit-of-happiness.org/history-of-happiness/ellen-langer.
17. "Mindfulness in the Age of Complexity."
18. Pursuit of Happiness, "Ellen Langer."
19. "Mindfulness in the Age of Complexity"; Pursuit of Happiness, "Ellen Langer"; The On Being Project.
20. Ken Robinson, "Do Schools Kill Creativity?," TED Talk, February 2006, https://www.ted.com/talks/sir_ken_robinson_do_schools_kill_creativity?referrer=playlist-the_most_popular_ted_talks_of_all_time&autoplay=true.
21. Rachel Wells, "The Top 10 Skills to Put on a Resume in 2024, From Research," *Forbes*, March 27, 2024, https://www.forbes.com/sites/rachelwells/2024/03/26/the-top-10-skills-to-put-on-a-resume-in-2024-from-research/?sh=13b3bec45a78; World Economic Forum, "Future of Jobs Report 2023," May 2023, https://www3.weforum.org/docs/WEF_Future_of_Jobs_2023.pdf.
22. Amy C. Edmondson, *Right Kind of Wrong: The Science of Failing Well* (New York: Simon & Schuster, 2023); Amy Edmondson, "Right Kind of Wrong: The Science of Failing Well," *Next Big Idea Club,* October 2, 2023, https://nextbigideaclub.com/magazine/right-kind-wrong-science-failing-well-bookbite/45323.

Chapter 2

1. Caitlin E. Krause, *Mindful by Design: A Practical Guide for Cultivating Aware, Advancing, and Authentic Learning Experiences*, 15–16 (Dallas: Corwin, 2019).

2. Lisa Feldman Barrett, "Research Papers," n.d., https://lisafeldmanbarrett.com/academic-papers; Lisa Feldman Barrett, https://lisafeldmanbarrett.com; College of Science, "Lisa Feldman Barrett," Northeastern University, n.d., https://cos.northeastern.edu/people/lisa-barrett.
3. Mass General Research Institute, "Lisa Feldman Barrett, Ph.D.," Massachusetts General Hospital, n.d., https://researchers.mgh.harvard.edu/profile/7109394/Lisa-Feldman-Barrett.
4. Barrett, "Research Papers"; Mass General, "Lisa Feldman Barrett, Ph.D."
5. Lex Fridman, "Lisa Feldman Barrett: Counterintuitive Ideas About How the Brain Works," Lex Fridman Podcast #129, October 4, 2020, https://www.youtube.com/watch?v=NbdRIVCBqNI; Scott Barry Kaufman, "Best of Series: Surprising Truths About the Human Brain: Lisa Feldman Barrett," Podcast, November 9, 2023. https://scottbarrykaufman.com/podcast/best-of-series-surprising-truths-about-the-human-brain-lisa-feldman-barrett.
6. Patrick R. Steffen, Dawson Hedges, and Rebekka Matheson, "The Brain Is Adaptive not Triune: How the Brain Responds to Threat, Challenge, and Change," *Frontiers in Psychiatry* 13 (2022), https://doi.org/10.3389/fpsyt.2022.802606; Wikipedia contributors, "Triune Brain," Wikipedia, last edited May 2, 2024, https://en.wikipedia.org/wiki/Triune_brain; The Interaction Design Foundation, "The Concept of the 'Triune Brain'," blog, n.d., https://www.interaction-design.org/literature/article/the-concept-of-the-triune-brain; Nic Russell, "Debunking the Lizard Brain Myth: Understanding Our Emotions," October 18, 2023, https://www.linkedin.com/pulse/debunking-lizard-brain-myth-understanding-our-emotions-nic-russell.
7. Steffen et al., "The Brain Is Adaptive Not Triune"; Gregg Henriques, "What Is the Triune Mind?" *Psychology Today,* March 23, 2023, https://www.psychologytoday.com/us/blog/theory-of-knowledge/202303/what-is-the-triune-mind; Gerald Wiest, "Neural and Mental Hierarchies," *Frontiers in Psychology* 3 (2012), https://doi.org/10.3389/fpsyg.2012.00516; Sarah McKay, "Rethinking the Reptilian Brain," Dr. Sarah McKay, June 24, 2020, https://drsarahmckay.com/rethinking-the-reptilian-brain; Therapy in a Nutshell, "The Triune Brain," YouTube, 2018.
8. McKay, "Rethinking the Reptilian Brain."

9. Barrett, "Research Papers"; https://lisafeldmanbarrett.com; College of Science, Lisa Feldman Barrett: Mass General Research Institute, "Lisa Feldman Barrett"; Molly Callahan, "It's Time to Correct Neuroscience Myths," Northeastern University College of Science, April 18, 2019, https://cos.northeastern.edu/news/its-time-to-correct-neuroscience-myths.

10. Dan Siegel, https://drdansiegel.com.

Chapter 3

1. Krista Tippett, "Dacher Keltner—the Thrilling New Science of Awe," *On Being with Krista Tippett,* February 2, 2023, https://onbeing.org/programs/dacher-keltner-the-thrilling-new-science-of-awe.

2. Dacher Keltner, "Strengthen Your Leadership with the Science of Awe," Knowledge at Wharton, August 21, 2023, https://knowledge.wharton.upenn.edu/article/strengthen-your-leadership-with-the-science-of-awe.

3. Keltner, "Strengthen Your Leadership with the Science of Awe"; Mary Pipher, "Orchestrating Wonder," *Psychotherapy Networker,* March/April 2023, https://www.psychotherapynetworker.org/article/orchestrating-wonder.

4. Keltner, "Strengthen Your Leadership with the Science of Awe."

5. Tippett, "Dacher Keltner."

6. Pipher, "Orchestrating Wonder."

7. Parts of this section are from Caitlin Krause, *Designing Wonder*, 2020, with my permission, https://caitlinkrause.gumroad.com/l/IxbVJ.

8. Søren Kierkegaard, *Kierkegaard's Writings, Volume 22: The Point of View*, Howard V. Hong and Edna H. Hong trans. (Princeton, NJ: Princeton University Press, 2009).

9. Cal Newport, *Deep Work: Rules for Focused Success in a Distracted World* (Hachette UK: Grand Central Publishing, 2016).

10. Ana Sandoiu, "How Brain Waves Enable Creative Thinking," *Medical News Today,* December 12, 2018, https://www.medicalnewstoday.com/articles/323956.

11. Maria Popova, "The Overview Effect and the Psychology of Cosmic Awe," *The Marginalian,* July 2, 2022, https://www.themarginalian.org/2012/12/18/the-overview-effect-and-the-psychology-of-cosmic-awe.

12. Houston We Have a Podcast, "The Overview Effect," NASA, August 30, 2019, https://www.nasa.gov/podcasts/houston-we-have-a-podcast/the-overview-effect.
13. Kiona N. Smith, "What Yuri Gagarin Saw from Orbit Changed Him Forever," *Forbes,* April 12, 2021, https://www.forbes.com/sites/kionasmith/2021/04/12/what-yuri-gagarin-saw-from-orbit-changed-him-forever/?sh=6c51eb2031f4.
14. Caitlin Krause, "Virtual Reality Talk on the Moon: Mindfulness Memory Palace by Caitlin Krause," YouTube, April 6, 2021, https://www.youtube.com/watch?v=OIBvPRbo2bc.
15. Parker J. Palmer, *A Hidden Wholeness: The Journey Toward an Undivided Life* (Wiley, 2022).

Chapter 4

1. Ogi Djuraskovic, "Big Data Statistics 2023: How Much Data Is in the World?," *FirstSiteGuide* (blog), October 4, 2023, https://firstsiteguide.com/big-data-stats; Kevin Bartley, "Big Data Statistics: How Much Data Is There in the World?" *Rivery* (blog), August 27, 2023, https://rivery.io/blog/big-data-statistics-how-much-data-is-there-in-the-world.
2. Michael Dimock, "Defining Generations: Where Millennials End and Generation Z Begins," Pew Research Center, January 17, 2019, https://www.pewresearch.org/short-reads/2019/01/17/where-millennials-end-and-generation-z-begins.
3. Susie Demarinis, "Loneliness at Epidemic Levels in America," *Explore* 16, no. 5 (2020): 278–79. https://doi.org/10.1016/j.explore.2020.06.008; Jean M. Twenge, Jonathan Haidt, Andrew B. Blake, Cooper McAllister, Hannah Lemon, and Astrid Le Roy, "Worldwide Increases in Adolescent Loneliness," *Journal of Adolescence* 93, no. 1 (2021): 257–269, https://doi.org/10.1016/j.adolescence.2021.06.006; Jean M. Twenge, Brian H. Spitzberg, and W. Keith Campbell, "Less In-person Social Interaction with Peers Among U.S. adolescents in the 21st Century and Links to Loneliness," *Journal of Social and Personal Relationships* 36, no. 6 (2019): 1892–1913. https://doi.org/10.1177/0265407519836170.
4. Matt Richtel, "'It's Life or Death': The Mental Health Crisis Among U.S. Teens," *New York Times*, May 3, 2022, https://www.nytimes.com/2022/04/23/health/mental-health-crisis-teens.html.

5. Mike Stobbe and the Associated Press, "Gen Z Homicides Hit 25-year High During COVID—and the Suicide Rate Was the Worst in Over 50Years, CDC Study Says," *Fortune Well*, June 15, 2023, https://fortune.com/well/2023/06/15/gen-z-mental-health-crisis-suicide-homicide-rates-cdc-study.
6. Gonzalo Martinez-Ales et al., "Why Are Suicide Rates Increasing in the United States? Towards a Multilevel Reimagination of Suicide Prevention," *Current Topics in Behavioral Neurosciences* 46 (2020): 1–23, https://doi.org/10.1007/7854_2020_158.
7. Devashru Patel et al., "Predicting State Level Suicide Fatalities in the United States with Realtime Data and Machine Learning," *NPJ Mental Health Research* 3 no. 1 (2024). https://doi.org/10.1038/s44184-023-00045-8.
8. The Jed Foundation, "Mental Health and Suicide Statistics," n.d., https://jedfoundation.org/mental-health-and-suicide-statistics; David D. Luxton, Jennifer D. June, and Jonathan M. Fairall, "Social Media and Suicide: A Public Health Perspective," *American Journal of Public Health* 102, no. S2 (2012): S195–S200. https://doi.org/10.2105/ajph.2011.300608.
9. Luxton, June, and Fairall, "Social Media and Suicide."
10. Gonzalo Martinez-Ales et al., "Why Are Suicide Rates Increasing in the United States?"
11. The Oboloo Team, "Hard Savings vs Soft Savings—What's the Difference?" *Oboloo* (blog), March 28, 2023, https://oboloo.com/blog/hard-savings-vs-soft-savings-whats-the-difference.
12. Jessica Walrack, "What Is Soft Saving and Do You Need to Be Doing It?" *U.S. News & World Report*, June 16, 2023, https://money.usnews.com/money/personal-finance/family-finance/articles/what-is-soft-saving; Alisa Wolfson, "Forget FIRE, Gen Z Is 'Soft Saving.' And Financial Advisers Have (ahem) a lot of Thoughts . . . ," *Market Watch*, November 2, 2023, https://www.marketwatch.com/picks/forget-fire-gen-z-is-soft-saving-and-financial-advisers-have-ahem-a-lot-of-thoughts-59825b31.
13. Walrack, "What Is Soft Saving and Do You Need to Be Doing It?"; Wolfson, "Forget FIRE, Gen Z Is 'Soft Saving.'"
14. Walrack, "What Is Soft Saving and Do You Need to Be Doing It?"; Liliana Hall, "Gen Z and 'Soft Saving': How I Balance My Financial and

Mental Health," CNET Money, January 22, 2024, https://www.cnet.com/personal-finance/banking/advice/gen-z-soft-saving.

15. Jamieson Webster, "Teenagers Are Telling Us That Something Is Wrong with America," *New York Times*, October 11, 2022, https://www.nytimes.com/2022/10/11/opinion/teenagers-mental-health-america.html.
16. James Clear, "How Habits Shape Your Health, Happiness, and Wealth," *James Clear* (blog), February 4, 2020, https://jamesclear.com/lewins-equation.
17. Michael Easter, https://gabriellereece.com/michael-easter-digital-media-and-our-psychology/.
18. Michael Easter, https://www.everand.com/podcast/517478593/The-Comfort-Crisis-With-Michael-Easter-Michael-Easter-is-a-leading-voice-on-how-humans-can-integrate-modern-science-and-evolutionary-wisdom-for-impr OR [ii] https://eastermichael.com.
19. Vivek H. Murthy, "Our Epidemic of Loneliness and Isolation: The U.S. Surgeon General's Advisory on the Healing Effects of Social Connection and Community," 2023, Office for the U.S. Surgeon General, https://www.hhs.gov/sites/default/files/surgeon-general-social-connection-advisory.pdf; Emily Harris, "Surgeon General Offers Strategy to Tackle Epidemic of Loneliness," *JAMA* 329, no. 21 (2023): 1818. https://doi.org/10.1001/jama.2023.8662.
20. Office of the Assistant Secretary for Health (OASH), "New Surgeon General Advisory Raises Alarm About the Devastating Impact of the Epidemic of Loneliness and Isolation in the United States," news release, HHS.Gov, May 3, 2023, https://www.hhs.gov/about/news/2023/05/03/new-surgeon-general-advisory-raises-alarm-about-devastating-impact-epidemic-loneliness-isolation-united-states.html; Bryan Robinson, "U.S. Surgeon General Cites Loneliness as Serious Mental Health Hazard in New Report," *Forbes*, September 12, 2023, https://www.forbes.com/sites/bryanrobinson/2023/05/06/us-surgeon-general-cites-loneliness-as-serious-mental-health-hazard-in-new-report.
21. OASH, "New Surgeon General Advisory Raises Alarm About the Devastating Impact of the Epidemic of Loneliness and Isolation in the United States"; Janice Hopkins Tanne, "Epidemic of Loneliness Threatens Public Health, Says US Surgeon General," *BMJ* 381 (2023): 1017, https://doi.org/10.1136/bmj.p1017.

Chapter 5

1. Sherry Turkle, "Connected, but Alone?," TED, February 2012, https://www.ted.com/talks/sherry_turkle_connected_but_alone.
2. The Wellbeing Project, "Chip Conley," n.d., https://wellbeing-project.org/chip-conley.
3. Maggie Appleton, "Ambient Co-presence," January 2024, https://maggieappleton.com/ambient-copresence.
4. Chuyue Ou and Zhongxuan Lin, "Co-presence, Dysco-presence, and Disco-presence: Navigating WeChat in Chinese Acquaintance Networks," *New Media & Society* (2023), https://doi.org/10.1177/14614448231168566.
5. Hacker News, "Ambient Co-Presence," February 2024, https://news.ycombinator.com/item?id=38824886.
6. Hacker News, "Ambient Co-Presence."
7. Ana Levordashka and Sonja Utz, "Ambient Awareness: From Random Noise to Digital Closeness in Online Social Networks," *Computers in Human Behavior* 60 (2016): 147–54, https://doi.org/10.1016/j.chb.2016.02.037.
8. Guy Winch, "My Story," n.d., https://www.guywinch.com/about; All American Speakers Bureau, "Guy Winch," n.d., https://www.allamericanspeakers.com/celebritytalentbios/Guy+Winch/393790.
9. Guy Winch, "Video Games Can Boost Emotional Health and Reduce Loneliness," *Psychology Today* (December 6, 2019), https://www.psychologytoday.com/us/blog/the-squeaky-wheel/201912/video-games-can-boost-emotional-health-and-reduce-loneliness.

Chapter 6

1. https://pz.harvard.edu/projects/center-for-digital-thriving
2. Harvard Graduate School of Education, "Emily Weinstein," n.d., https://www.gse.harvard.edu/directory/faculty/emily-weinstein.
3. Nell Shapiro, "Deliberately Minimal: Teens & Technology With Dr. Emily Weinstein," *Kibou* (blog), August 16, 2022, https://kiboubag.com/blogs/kibou-blog/deliberately-minimal-teens-technology-with-dr-emily-weinstein.

4. Kendra Cherry, "A Comprehensive Guide to the Bronfenbrenner Ecological Model," Verywell Mind, August 16, 2023, https://www.verywellmind.com/bronfenbrenner-ecological-model-7643403.
5. Bronfenbrenner Center for Translational Research, "Urie Bronfenbrenner," Cornell Human Ecology, n.d., https://bctr.cornell.edu/about-us/urie-bronfenbrenner; Practical Psychology, "Urie Bronfenbrenner Biography–Contributions to Psychology," March 2021, https://practicalpie.com/urie-bronfenbrenner.
6. Bronfenbrenner Center for Translational Research, "Urie Bronfenbrenner"; Practical Psychology, "Urie Bronfenbrenner Biography–Contributions to Psychology."
7. Bronfenbrenner Center for Translational Research, "Urie Bronfenbrenner."
8. Cherry, "A Comprehensive Guide to the Bronfenbrenner Ecological Model"; Paul Main, "Bronfenbrenner's Ecological Model," Structural Learning, May 5, 2023, https://www.structural-learning.com/post/bronfenbrenners-ecological-model.
9. Common Sense Education, "Rings of Responsibility," YouTube, August 10, 2018, https://www.youtube.com/watch?v=fQSnzrB5bso.
10. Emily Weinstein and Carrie James, *Behind Their Screens: What Teens are Facing (and Adults Are Missing)* (Cambridge, MA: MIT Press, 2022).
11. Learning Planet Institute, https://www.learningplanetinstitute.org/en.

Chapter 7

1. "Professor Emeritus Seymour Papert, Pioneer of Constructionist Learning, Dies at 88," *MIT News*, August 1, 2016.
2. "Professor Emeritus Seymour Papert, Pioneer of Constructionist Learning, Dies at 88," *MIT News*; Nicole Ellison, "Seymour Papert," *Encyclopedia Britannica*, last updated May 8, 2024, https://www.britannica.com/biography/Seymour-Papert.
3. Peter Mark Howell, "Disruptive Game Design," doctoral dissertation, University of Portsmouth, November 2015, https://researchportal.port.ac.uk/en/studentTheses/disruptive-game-design.
4. r/gamedesign, "What Are Some Examples of Bad Game Design?," Reddit, n.d., https://www.reddit.com/r/gamedesign/comments/2rfg6c/what_are_some_prime_examples_of_bad_game_design/?rdt=53444.

5. sipoCodes, "The Intersection of Frontend Development and Manipulative Psychology in Game Design: Dangers, Ethics, and Solutions in the Age of Artificial Intelligence," *Medium*, January 15, 2023, https://bootcamp.uxdesign.cc/the-intersection-of-frontend-development-and-manipulative-psychology-in-game-design-dangers-6ee65f502ce?gi=5cd978448572.
6. sipoCodes, "The Intersection of Frontend Development and Manipulative Psychology in Game Design: Dangers, Ethics, and Solutions in the Age of Artificial Intelligence."
7. sipoCodes, "The Intersection of Frontend Development and Manipulative Psychology in Game Design."
8. Karl Purcell, "Behavioral Game Design: 7 Lessons from Behavioral Science to Help Change User Behavior," Irrational Labs, August 3, 2022, https://irrationallabs.com/blog/behavioral-game-design-7-lessons-from-behavioral-science-to-help-change-user-behavior.
9. Zukun Lyu, "The Manipulative Techniques of Online Games in Data Collection," Stratheia, November 7, 2023, https://stratheia.com/the-manipulative-techniques-of-online-games-in-data-collection/?amp=1.
10. Charlene Jennett et al., "Measuring and Defining the Experience of Immersion in Games," University College London Interaction Centre, n.d., https://citeseerx.ist.psu.edu/document?doi=748020e60efbfdbef817101242c88e455dce0bef&repid=rep1&type=pdf.
11. Brigid Mary Costello, "Rhythmic Entrainment in Games," in *Lecture Notes in Computer Science*, 2016, 220–30, https://doi.org/10.1007/978-3-319-45841-0_21.
12. Bulbagarden, "Entrainment (Move)," n.d., https://bulbapedia.bulbagarden.net/wiki/Entrainment_(move).
13. The UnLonely Project, "Creative Expression Approach," The Foundation for Art & Healing, n.d., https://www.artandhealing.org/creative-expression-approach.
14. Elena Lewis, "The Transformative Power of Creative Arts: Nurturing Mental Health Through Expression," *GPS Lansing* (blog), July 19, 2023, https://www.guidetopersonalsolutions.com/post/the-transformative-power-of-creative-arts-nurturing-mental-health-through-expression; Jefferson Center, "How Flexing Your Creativity Can Benefit Your Health," *Mental Health Matters* (blog), n.d., https://www.jcmh.org/how-flexing-your-creativity-can-benefit-your-health.

15. The UnLonely Project, "Creative Expression Approach."
16. Arielle Schwartz, "Embodiment in Somatic Psychology," Center for Resilience Informed Therapy, March 25, 2017, https://drarielleschwartz.com/embodiment-in-somatic-psychology-dr-arielle-schwartz.
17. Schwartz, "Embodiment in Somatic Psychology."
18. Serena Lee-Cultura and Michail Giannakos, "Embodied Interaction and Spatial Skills: A Systematic Review of Empirical Studies," *Interacting with Computers* 32, no. 4 (2020): 331–366. https://doi.org/10.1093/iwcomp/iwaa023.
19. Kiin Knowledge Base, "What Is 'Embodiment' in Virtual Reality?" *Kiin* (blog), June 24, 2023, https://kiin.tech/blog_kiin/what-is-embodiment-in-virtual-reality.
20. Kiin Knowledge Base, "What Is 'Embodiment' in Virtual Reality?"
21. Shona Erskine, "The Integration of Somatics as an Essential Component of Aesthetic Dance Education," AusDance, July 1, 2004, https://ausdance.org.au/articles/details/the-integration-of-somatics-as-an-essential-component-of-aesthetic-dance-ed.
22. Richard Shusterman, "Somaesthetics," Interaction Design Foundation, January 1, 2014, https://www.interaction-design.org/literature/book/the-encyclopedia-of-human-computer-interaction-2nd-ed/somaesthetics.
23. Stefan Brunnhuber and Andreas Michalsen, "Psychosomatik Und Mind–Body-Medizin: Integrative, Komplementäre Oder Alternative Disziplinen? Ein Entwicklungslogisches Argument," *Research in Contemporary Medicine* 19, no. 2 (2012): 86–92, https://doi.org/10.1159/000338537; Tobias Esch and George B Stefano, "The BERN Framework of Mind–Body Medicine: Integrating Self-Care, Health Promotion, Resilience, and Applied Neuroscience," *Frontiers in Integrative Neuroscience* 16 (2022), https://doi.org/10.3389/fnint.2022.913573; Maureen Salamon, "What Is Somatic Therapy?" *Harvard Health* (blog), July 7, 2023, https://www.health.harvard.edu/blog/what-is-somatic-therapy-202307072951; Mount Sinai Health System, "Mind–body Medicine," n.d., https://www.mountsinai.org/health-library/treatment/mind-body-medicine.
24. Celia Hodent, "Cognitive Psychology Applied to User Experience in Video Games," *Celia Hodent* (blog), March 4, 2016, https://celiahodent

.com/cognitive-psychology-applied-to-user-experience-in-video-games; Celia Hodent, "The Gamer's Brain: How Neuroscience and UX Can Impact Design (GDC15)," *Celia Hodent* (blog), March 31, 2015, https://celiahodent.com/video-game-ux-psychology.

25. Hodent, "The Gamer's Brain."

26. Raph Koster, "The Healing Game," Raph Koster's Website, March 2, 2006, https://www.raphkoster.com/2006/03/02/the-healing-game.

27. Raph Koster, "Moore's Wall: Technology Advances and Online Game Design," n.d., https://www.raphkoster.com/gaming/moore.shtml.

28. Koster, "Moore's Wall."

29. Koster, "The Healing Game."

30. Ben Dickson, "What Is Spatial Computing? A Basic Explainer," *PC Mag,* January 19, 2024, https://www.pcmag.com/how-to/what-is-spatial-computing-a-basic-explainer.

31. Internova Travel Group, "Best Immersive Wellness Destinations," n.d., https://internovatravel.com/category/hotels/immersive-wellness-destinations.

32. Diana Olynick, *Interfaceless: Conscious Design for Spatial Computing with Generative AI (Design Thinking)* (Cambridge, MA: MIT Press, 2024).

33. Olynick, *Interfaceless.*

34. Horacio Torrendell, "Top 9 Examples of Spatial Computing," *Treeview Studio Blog*, October 13, 2023, https://treeview.studio/blog/top-examples-of-spatial-computing.

35. Torrendell, "Top 9 Examples of Spatial Computing."

36. "Can Spatial Computing Reshape Health Care?" Stambol, January 21, 2023, https://www.stambol.com/2019/01/21/can-spatial-computing-reshape-health-care.

37. "Can Spatial Computing Reshape Health Care?" Stambol.

38. "Haven Creates Scalable Access to Mental Health Experiences Using Immersive Technology," Immersive Learning News, November 15, 2023, https://www.immersivelearning.news/2023/11/15/haven-creates-scalable-access-to-mental-health-experiences-using-immersive-technology.

39. "Haven Creates Scalable Access to Mental Health Experiences Using Immersive Technology," Immersive Learning News.

Chapter 8

1. Caitlin Krause, "The Seven S's of the Metaverse," *Medium*, July 5, 2022, https://caitlinkrause.medium.com/the-seven-ss-of-the-metaverse-23a8bdd318b6.
2. James Jensen, "How James Jensen Founded JUMP by Limitless Flight," *Utah Business*, May 18, 2023, https://www.utahbusiness.com/how-james-jensen-founded-jump-by-limitless-flight-base-jumping; Chemain Evans, "James Jensen, Creator of the VOID, Creator of JUMP, World-Building Extraordinaire," *Lemonade Stand* (blog), December 12, 2020, https://blog.lemonadestand.org/james-jensen-creator-of-the-void-creator-of-jump-world-building-extraordinaire; Jacqueline Mumford, "James Jensen Closes the VOID and Opens JUMP," *Utah Business*, April 7, 2021, https://www.utahbusiness.com/james-jensen-closes-the-void-and-opens-jump.

Chapter 9

1. Competition & Markets Authority, "Music and Streaming," Crown, November 29, 2022, https://assets.publishing.service.gov.uk/media/6384f43ee90e077898ccb48e/Music_and_streaming_final_report.pdf.
2. Jannik Lindner, "Must-Know Music Consumption Statistics [Current Data]," Gitnux, December 16, 2023, https://gitnux.org/music-consumption-statistics.
3. Lindner, "Must-Know Music."
4. Rubin Museum of Art, "The World Is Sound," 2017, https://rubinmuseum.org/events/exhibitions/the-world-is-sound.
5. Jas Brooks and Pedro Lopes, "Smell & Paste: Low-Fidelity Prototyping for Olfactory Experiences," *Proceedings of the 2023 CHI Conference on Human Factors in Computing Systems (CHI '23)*, 2023, https://lab.plopes.org/published/2023-CHI-SmellAndPaste.pdf.
6. EVRYTHNG and Avery Dennison, *Digital Emotional Intelligence: A Customer Experience Framework to Understand and Anticipate Human Emotion Across Physical and Digital Channels* (2017), https://rbis.averydennison.com/content/dam/averydennison/rbis/global/apparel/Documents/Avery-Dennison-Digital-Emotional-Intelligence.pdf.

7. Amy Rigby, "Science-backed Productivity Playlists to Help You Dive into Deep Work—Work Life by Atlassian," *Work Life* blog, Atlassian, December 29, 2022, https://www.atlassian.com/blog/productivity/science-backed-productivity-playlists; Júlia Araújo, "The 4 Type of Music for Productivity, According to Science," *Akiflow* (blog), July 11, 2022, https://akiflow.com/blog/music-for-productivity-reduce-stress-and-improve-focus.
8. Music Health, "Music and the Vagus Nerve: How Music Affects the Nervous System and Mental Health," n.d., https://www.musichealth.ai/blog/music-and-the-vagus-nerve.
9. Chris Marrano, "2 Ways Music Can Affect the Vagus Nerve and Its Benefits in FL," *Neuvana* (blog), October 14, 2020, https://neuvanalife.com/blogs/blog/2-ways-music-can-affect-the-vagus-nerve-and-its-benefits-in-fl.
10. Alex Reijnierse, "Vagus Nerve Stimulation Music: Advanced Sound Therapy," *Paleo Stress Management* (blog), May 1, 2024, https://paleostressmanagement.com/vagus-nerve-stimulation-music.

Chapter 10

1. Canadian Agency for Drugs and Technologies in Health, *Neurofeedback and Biofeedback for Mood and Anxiety Disorders: A Review of the Clinical Evidence and Guidelines—an Update* (Ottawa, Canada: Canadian Agency for Drugs and Technology in Health, 2014), https://www.ncbi.nlm.nih.gov/books/NBK253820.
2. Nuno Gomes et al., "A Survey on Wearable Sensors for Mental Health Monitoring," *Sensors* 23, no. 3 (2023): 1330, https://doi.org/10.3390/s23031330.
3. Corinne Iozzio, "Stressed? The Latest in Wearables Could Help Keep You Calm," *Smithsonian Magazine*, July 8, 2014, https://www.smithsonianmag.com/innovation/wearable-tracker-helps-you-keep-calm-180951981.
4. WHOOP, "Everything You Need to Know About Heart Rate Variability (HRV)," WHOOP, August 11, 2021, https://www.whoop.com/us/en/thelocker/heart-rate-variability-hrv.

5. Cleveland Clinic, "Heart Rate Variability (HRV)," n.d., https://my.clevelandclinic.org/health/symptoms/21773-heart-rate-variability-hrv.
6. Reena Tiwari et al., "Analysis of Heart Rate Variability and Implication of Different Factors on Heart Rate Variability," *Current Cardiology Review* 17, no. 5 (2021).
7. Kinza Yasar and Ivy Wigmore, "Wearable Technology," *Mobile Computing* (2022); Reham Alhejaili and Akram Alomainy, "The Use of Wearable Technology in Providing Assistive Solutions for Mental Well-Being," *Sensors* 23, no. 17 (2023): 7378, https://doi.org/10.3390/s23177378; Katie Navarra, "New Uses for Wearable Devices in the Workplace," *SHRM*, March 14, 2022, https://www.shrm.org/topics-tools/news/technology/new-uses-wearable-devices-workplace; "Using Wearable Technology to Enhance Corporate Wellness Programs," *Corporate Wellness Magazine,* n.d., https://www.corporatewellnessmagazine.com/article/using-wearable-technology-to-enhance-corporate-wellness-programs?28ee73ed_page=6.
8. Seunggyu Lee, Hyeson Kim, and Mi Jin Park, "Current Advances in Wearable Devices and Their Sensors in Patients with Depression," *Frontiers in Psychiatry* 12 (2021). https://doi.org/10.3389/fpsyt.2021.672347.
9. Manoush Zomorodi et al., "Over 20,000 Joined the NPR/Columbia Study to Move Throughout the Day. Did It Work?," *NPR*, November 7, 2023, https://www.npr.org/2023/11/07/1200611641/changing-our-sedentary-screen-filled-habits; "Body Electric," *NPR*, n.d., https://www.npr.org/series/1199526213/body-electric.
10. Zomorodi et al., "Over 20,000 Joined the NPR/Columbia Study to Move Throughout the Day."
11. Sarah Meyer Tapia, "Running Naked: A Thematic Analysis of the Lived Experience Running With and Without Technology," unpublished doctoral dissertation, California Institute of Integral Studies, 2023.

Chapter 11

1. "The State of AI in 2023: Generative AI's Breakout Year," McKinsey & Company, August 1, 2023, https://www.mckinsey.com/capabilities/quantumblack/our-insights/the-state-of-ai-in-2023-generative-ais-breakout-year.

2. Mauro Cazzaniga et al., "Gen-AI: Artificial Intelligence and the Future of Work," IMF Staff Discussion Note, https://www.imf.org/-/media/Files/Publications/SDN/2024/English/SDNEA2024001.ashx.
3. Nestor Maslej et al., "Artificial Intelligence Index Report 2024," Institute for Human-Centered Artificial Intelligence, Stanford University, April 2024, https://aiindex.stanford.edu/report.
4. Nathan E. Sanders and Bruce Schneier, "Let's Not Make the Same Mistakes with AI That We Made with Social Media," *MIT Technology Review*, March 13, 2024, https://www.technologyreview.com/2024/03/13/1089729/lets-not-make-the-same-mistakes-with-ai-that-we-made-with-social-media.
5. Tristan Harris and Aza Raskin (hosts), "The AI Dilemma," *Your Undivided Attention* (podcast), March 24, 2023, https://www.humanetech.com/podcast/the-ai-dilemma.
6. Tristan Harris and Aza Raskin (hosts), "The Three Rules of Humane Tech," *Your Undivided Attention* (podcast), April 6, 2023, https://www.humanetech.com/podcast/the-three-rules-of-humane-tech.
7. Tristan Harris, "We Think in 3D. Social Media Should, Too," *TIME*, January 12, 2023, https://time.com/6246077/social-media-3d.

Future Sight: Envisioning a Future of Digital Thriving

1. Hannah Featherman, "Tree Profile: Aspen—So Much More Than a Tree," National Forest Foundation, n.d., https://www.nationalforests.org/blog/tree-profile-aspen-so-much-more-than-a-tree.
2. Priyanka B. Carr and Gregory M. Walton, "Cues of Working Together Fuel Intrinsic Motivation," *Journal of Experimental Social Psychology* 53 (2014): 169–84, https://doi.org/10.1016/j.jesp.2014.03.015.

Acknowledgments

This book is written on three levels: the Human level, where we exist as "breathers on land"; the Deep Sea level, where we take a breath, dive deep, swim buoyant, and form new understandings; and the Space level, where we're uninhibited by gravity, no up and down, only open and out, and we are surrounded by stars.

Light reaches us on all three levels, and this book has allowed me to invite everyone to exist on those three levels at once, seeing from different views, experiencing Freedom, and aiming, always, to keep our sights on the uplifting, animating feelings of Wonder and Imagination. . . . We use a human sensibility and a transcendence, all at once. Ultimately, we rise to a different level of being through greater understandings and feelings of agency. There are practical angles, meeting us where we're at and solving acute needs for right now, and broader angles, the larger sense of Creativity that is the essence of buoyant Flow, guiding us to a future that has a chance to be expansive, expressive, and joyfully connected.

I am grateful to all who offered support, research, interviews, advice, time, attention, care, and compassion. I would like to name people who served in many different ways to help bring this book to life:

Kathleen Krause
Sarah Meyer Tapia
Donald McKendrick
Marc Prensky

Atanas Bakalov
Brad Rochefort
Emily Krause
Jude Shannon
Andy Fidel
Sam Perry
Nina Hersher
John Vitale
Tom Furness
Mark Billinghurst
Cody Ray
Ryan Douglas
Julia Scott
Norm Adams
Steve Carnevale
Cloud Force
Rodney Mullen

Thank you to the editing team, including Victoria Savanh, for recognizing the necessity of these messages from the beginning, and to Julie Kerr for editing this book with your capable care.

Luna, thank you for being the best "poetry pup" I could wish for.

About the Author

Caitlin Krause is an experienced designer, global speaker, and author who focuses on the intersection of creativity, technology, innovation, and wellbeing. She founded the consultancy MindWise in 2015 and teaches about digital wellbeing and extended reality (XR) at Stanford University. Author of *Designing Wonder: Leading Transformative Experiences in XR, Mindful by Design,* and *Digital Satori*, Caitlin has advised global organizations including TED, Google, the US Air Force, Meta, Oracle, Evernote, University of San Francisco, ETH Zürich, and the US State Department. She has created and run numerous collaborative experiences in social XR, fusing presence, storytelling, meditation, AI, advanced technology, and social and emotional intelligence. She helps to create immersive therapeutic applications that engage the senses for a full experience. She holds an MFA from Lesley University and a BA from Duke University. Caitlin helps leaders and teams navigate complexity and change in the future of work, prioritizing wellbeing, collaboration, curiosity, and imagination through wonder and awe.

Index

Page numbers followed by *f* refer to figures.